Stay Me With Flagons

Maurice Healy

Published 2021 by Académie du Vin Library Ltd
academieduvinlibrary.com
Founders: Steven Spurrier and Simon McMurtrie

Publishers: Simon McMurtrie and Hermione Ireland
Editor: Susan Keevil
Art Director: Tim Foster
Typesetting: Cheshire Typesetting

ISBN: 978-1-913141-23-3
Printed and bound in the UK

First published 1940

Stay Me With Flagons

Maurice Healy

A Book About Wine and Other Things

Introduced by Fiona Morrison MW

ACADEMIE DU VIN LIBRARY • CLASSIC EDITIONS

Académie du Vin Library Classic Editions

In the Vine Country by Edith Somerville and Martin Ross – (1893)

Wayward Tendrils of the Vine by Ian Maxwell Campbell – (1948)

A Contemplation of Wine by H Warner Allen – (1951)

academieduvinlibrary.com

Our Classic Editions have each been selected for the way they capture the vineyards, wines, people and culture of vintages gone by and for the engaging writing style in which they're presented. To preserve the authors' unique style, the language of these books has been left unedited and may not reflect the values we adhere to in our commentaries today.

Académie du Vin Library Classic Editions

Our Académie du Vin Library 'Classic Editions' lie at the very heart of the books we publish. When Steven Spurrier and I first talked about a new publishing imprint over lunch at our club (where else!), it was the older wine books we thought of first. These islands have been the heartland of wine appreciation, it seems, since Chaucer's time and there is much literature to draw from: wine writing that is elegant, informative, inspiring, often eccentric and frequently witty. Some of it feels dated now but that's part of its charm. What a shame it would be, we said to one another, if the words that inspired our own wine careers were lost to today's reader. Indeed, they shouldn't be! So while we at the Académie celebrate new authors, new adventures and new perspectives on our pages (and always will), we are now very proud to devote a new series of 'time capsule' titles to the wine writers who handed the job – if you can call it that – over to us.

I have shelves of these books, and quite a few favourites. Here, we welcome back Maurice Healy (1887–1943), the Irish lawyer, author and poet who won the hearts of readers and friends with his free-flowing wit and wine observation. I must admit, I've always admired him.

Hugh Johnson

About the Author

Maurice Healy was born in Cork, Ireland, in 1887. He trained as a lawyer at University College Dublin and was called to the Irish Bar in 1910, then the English Bar in 1914. His experiences of World War I (on the Western Front and at Gallipoli) gave him a horror of violence and he opted to practise in England rather than remain in Ireland amid its wars of independence (1919–21). In his memoir of Healy (on pages 9–14 of this book), The Right Hon Sir Norman Birkett recalls: '…to have known him is to have known the Bar at its very best.' But Healy's interests extended way beyond the law, and as Birkett explains, 'he found the tang and savour of his many-sided life too attractive to be sacrificed'.

Poetry, politics, broadcasting and the pursuit of lively conversation all played a part, but highest of all amongst Healy's affections, perhaps, were 'matters affecting the table' – wine in particular. Claret and conversation went hand in hand, and in *Stay Me With Flagons* Healy not only eloquently sets out his extensive wine knowledge but recalls in vivid detail bottles opened while dining with fellow wine writers including the luminaries André Simon, Ian Maxwell Campbell and George Saintsbury.

Maurice Healy died in 1943. He was, sadly, never to realize the affection instilled by this book in the generation of wine-lovers that followed him. This is its seventh printing.

Introduction
Fiona Morrison MW

What a wonderfully personal, poetic account of wine this is by a brilliant Irish raconteur! The title of the book comes from a verse of the ancient text, the *Canticle of Canticles*, which speaks of wine as an honourable gift of God. God and wine are two key characters in this book, the third being the author himself, Maurice Healy. Born at the end of the 19th century in Ireland, he was a successful barrister, who undoubtedly honed his way with words practising the law.

Healy grew up as a teetotaller under the influence of the Catholic Church but threw himself into the worship of wine with the energy of the converted when he finally discovered it, aged 30. His description of his first taste of wine, in the company of his equally rebellious parish priest, is one of the first of many vivid tableaux in the book. Sadly, he was to die at the age of only 55, shortly after this book was published, giving him only 25 intense years to indulge his passion.

This is essentially an amateur wine book, in both senses of the word. Healy is a passionate lover of wine whose story rollicks along with humour, passion, erudition and fun. And if his tales of the origins and history of events differ from our received wisdom, then we are all the more enlightened by his alternative version. Here and there editorial notes gently correcting or clarifying Healy's text are made by Ian Maxwell Campbell, whose own book, *Wayward Tendrils of the Vine*, also forms part of the Académie du Vin Library's Classic Editions series. Campbell's interjections are hilarious, rather like those of an indulgent schoolmaster trying to temper the unbridled enthusiasm of a gifted pupil.

A wonderful storyteller, Healy's pages resonate with the melodic lilt of his Irish brogue. There is a poetic language here that speaks of chivalry, of gentlemen's clubs and public schools, of the cricket pitch and of the grouse moor. Healy and his coterie of fellow wine lovers all seem to have written witty memoirs in praise of wine and the generosity of friends. Forgive him his prejudices, his chauvinistic views of women's palates, the obstinacy in his views and old-fashioned style, and you'll find much of the pleasure of this book comes from indulging him and letting his booming voice leap from the pages in joyful gusto.

Yet while the jollity will charm, we shouldn't ignore the seriousness of the history, the traditions, the culture of wine that Healy writes about so vividly. After reading this book, you will have a sense of richness, an extra layer of wine understanding that history always confers on us.

Of course, claret takes up most of the book's descriptions: 'It is *the* wine,' Healy avers at the beginning of three chapters dealing with the subject, in which he places Lafite at the top of the Bordeaux pinnacle; claret, sherry, port, champagne, burgundy, Rhône and hock follow closely after (his descriptions of German wines are particularly vivid and include the hope that 'the vineyards at the moment are not paying too dearly for Herr Hitler'). Other regions only have walk-on roles, and while his descriptions of how their wines are made are elegiac, they are not completely accurate. Healy's is a rather benign wine world where the art rather than the science of wine comes to the fore.

The stories are abundantly peppered with historical anecdotes bringing in Charlemagne, Nicolas Rolin, Mussolini, the Knights Templar, numerous popes and monarchs, and the Girondins – about whom an amazing story of daring and disaster in the limestone cellars of Saint Emilion ends with the words 'visit Saint Emilion with an ear for the tread of its tragic ghosts'.

The book was written during World War II, and it seems incredible that Healy wrote of such abundance (he was known for being somewhat 'portly') during a time of rationing and rarity. But many of us can warm to his words that the pleasure of wine exists as much in the excellence of the conversation it stimulates as its quality, and his booming exuberance in the pleasure of wine fills the pages. Nicknamed 'The Prattler', he would set the table on a roar, usually at his favourite drinking place, The Hind's Head at Bray. He was known to love Château Haut-Brion because its spelling could so nearly be Irish, 'O'Brian'.

We have to read these vinous words with unabashedly rose-tinted spectacles. They are much more than mere guides for the wine drinker – they are memoirs. Therein lies their charm. I find myself amused that in writing this introduction I am unconsciously slipping into his – the language of Evelyn Waugh or PG Wodehouse. Reading Healy's description of a magnificent bottle of Volnay Caillerets on page 167 puts us modern wine-writing hacks to shame. It expresses more about the wine and its taste than any modern-day tasting note or 100-point score.

This is an extremely personal book: Healy sets down his own opinions and judgements in a much franker way than any writer would do today. At the beginning of the book, he writes: 'We are going to have a grand time together' and he certainly delivers on his promise.

The book finishes with two of his poems: half witty doggerel, half lyrically poetic. They sum up the voice of a wine lover and raconteur I would love to have met.

STAY ME WITH FLAGONS

First published during the dismal wartime winter of 1940, this book met with a response which delighted and to some extent surprised its author. The complete destruction by enemy action of the standing type, allied with increasing shortage of paper and obedience to the austerities of the time, drove the book out of print, and for many years any stray copy has been a 'collector's' item.

Maurice Healy died in May, 1943, and Sir Norman Birkett pays tribute to his memory in the appreciation he contributes to this new edition, of which this is the fourth impression.

Very little that Healy wrote has been contradicted by time or events. But wines, like time, change and in order that such changes should be noted, Colonel Ian Campbell (an old friend of Maurice Healy's and on wine an authority of the *premier cru*) has interpolated a series of erudite and endearing annotations. These are printed in italics and appear in the body of the text.

DEDICATED TO

MY FRIEND

BARRY NEAME,

OF 'THE HIND'S HEAD,' BRAY,

Sometimes the Host of Princes,

Always the Prince of Hosts

CONTENTS

A MEMOIR

by

THE RIGHT HON. SIR NORMAN BIRKETT

THERE are certain rare and choice spirits about whom one feels that they should not be subject to the relentless laws of mortality which govern ordinary men. It seems wrong that they should ever die. On a sunny morning in June, when the English team comes out from the pavilion at Lord's, it still seems strange that Grace is not among them. It would be quite in accordance with the eternal fitness of things if that great bearded figure were still there, and that high-pitched voice could still be heard. And whenever a Bar Mess of the Midland Circuit is gathered at Aylesbury or Lincoln or where not, or whenever the 'Odd Volumes,' that unique dining club, meets in the atmosphere of fellowship that so distinguishes it, it not only seems strange, it seems quite unthinkable that Maurice Healy will not again set the table on a roar as he was wont to do. But Maurice is gone; and something of gaiety and brightness has gone too, leaving the world a very much poorer place for those of us who remain. It cannot be said of him as Motley said of William of Orange that 'when he died the little children cried in the streets,' but when Maurice died, many a man found himself echoing the words of Johnson on the death of Garrick—'I am disappointed by that stroke of death, which has eclipsed the gaiety of nations and impoverished the public stock of harmless pleasure.'

This new edition of 'Stay Me With Flagons' would have given Maurice much pleasure, and I am greatly honoured by the invitation to contribute a Memoir. But I must make plain the poverty of my equipment for one part of the task. So far as I can judge from the introductory pages of Maurice's own book on Claret, a lifetime of experience is needed to qualify for the exalted position of an expert on wine. Many books must be read in many languages; wine merchants must be induced to become personal friends; from time to time, visits must be paid to the vineyards themselves; wines must be tasted and compared; private and public dinners must receive considerable attention; and whilst all this, no doubt, gives intense pleasure in the doing of it, it is not to be denied that an inordinately busy life puts considerable difficulties in the way. I hope I may be forgiven, therefore, if I say that these delights have not been possible for me, so that I am quite unable to assess with any critical eye the undoubted merits of this book; but the printing of a new edition would seem, of itself, to set the seal of excellence upon it.

For myself, I must confess that the language of the experts occasionally seems to me to be a little forced and artificial, and if it is not plain blasphemy to say so, at times a trifle ridiculous. I suppose that Mr. James Thurber suffers at times from the same feelings, or else how could he have produced that drawing in *The New Yorker*, where the host on introducing his wine to the guests remarks: 'It's a naïve domestic Burgundy, without any breeding, but I think you'll be amused by its presumption.' This, no doubt, is satirical; but Maurice himself in the present book has a number of pretty conceits, as when he says in the chapter on Burgundy:

'When one thinks of Bertrand du Guesclin the image is as much of good manners and courtesy as of valour; so I shall borrow him as a type of Romanée Conti, ranking the others as his brothers-in-arms. For the suavity of these wines would not show to such perfection were it not for their valour and sturdiness. They are indeed princely wines, and the qualities that lurk in their depths are courage and fair dealing.' This language, as I say, always sounds to me a shade artificial, but I am sure the defect lies in me, and in my want of the true education; and it cannot obscure the profundity of the learning with which the book is filled.

But of the author of the book I should like to speak with warmth and with deep affection, with no reservations of any kind.

The bare facts of his life are fairly well known. He was born in Ireland in 1887 and was educated there. In 1910 he was called to the Irish Bar, and readers of his book *The Old Munster Circuit* will realize the profound effect upon the whole of his life made by these early experiences in the formative years. He found time to contest a parliamentary election, and to write a certain amount of poetry, some of the poems surviving in Anthologies to this day. In 1914 he was called to the English Bar and joined the Midland Circuit, where he soon became one of its best-loved members. After service in the war, where he won the Military Cross, he began his legal career, and acquired a considerable practice. He was made King's Counsel in 1931, and later a Bencher of the Inner Temple. In 1934, he published his book on *Claret and the White Wines of Bordeaux*, and in 1940 the first edition of this present work appeared. These dates cover a life of remarkable activity, but it is difficult to express and reveal the remarkable quality of the life itself.

I first saw Maurice when he was conducting a case at the Leicester Assizes. I had heard of his coming to the Circuit and was much attracted by his brogue, though it was slight as compared with Lord Carson, and I stayed to watch his performance. That night I dined with him at the Bar Mess, and it was plain to me that a new figure of great brilliance had come among us. Since that day I have fought many cases with him and against him, and have had the pleasure of hearing him in the Court of Criminal Appeal when I was sitting as a member of that Court. In whatever capacity he appeared he was the same delightful personality. He certainly possessed those princely qualities of valour and courtesy which he so much admired, and to have known him is to have known the Bar at its very best.

I have always thought that Maurice never reached the place in the profession to which his great and varied gifts seemed to entitle him, and why this should have been so, must always remain something of a mystery. He was endowed with all the qualities requisite for the highest achievements and yet never quite reached them. It may be that his richly dowered mind did not permit him to pursue the narrow path of the law with a single and undeviating purpose, and that he found the tang and savour of his many-sided life too attractive to be sacrificed. It may be that his sparkling sense of humour sometimes got out of his control and created the impression, quite wrongly, that he took things too lightly.

But however that may be, in one sphere he was quite supreme, and that was the sphere where his wonderful personality had full scope and force.

He was a man of the rarest kind of charm, spontaneous, natural, and unaffected, and a charm, moreover, that defies all attempts at analysis. Perhaps one or two of the elements can be stated. He possessed the most lovely speaking voice and the most winning and attractive smile, not inconsiderable gifts in the rough and tumble of this world. His every action was characterized by a thoughtfulness for others and a most scrupulous regard for their feelings. He had a kind of insight, perhaps the most precious of all the gifts in Heaven's treasury, from which sprang his wide understanding of all sorts and conditions of men, his ready sympathy, and his intense and unfailing loyalties to great causes and to his many friends. He was capable of intense anger on occasion, but this was reserved for injustice wherever it was displayed, whether in national or individual affairs. He was the perfect host and the perfect guest, and I knew him well in both capacities. One of my earliest and happiest memories of Maurice is when he was host. I had said to him that I thought that Alice Meynell's sonnet, 'Renouncement,' was one of the very greatest in the language, and had gone on to speak with admiration of all her work. There followed a luncheon party at Canuto's where Alice Meynell was the guest of honour so that her admirer might meet her face to face; and the charming deference Maurice paid to that great figure, and the way he made the least important of his guests feel that they were equally welcome, will always remain with me.

He was my guest on many occasions too. These were times, I am bound to say, when I felt a slight sense of anxiety because of his knowledge, and his growing reputation in all matters affecting the table. I had listened to his paper on 'Irish Wine' at a meeting of the 'Odd Volumes,' and such knowledge seemed to me quite fearsome, and demanded excessive care in the character of the food and drink provided. I consulted experts and produced something quite out of the ordinary, which I could not forbear from introducing with some such phrase as 'Here, my dear Maurice, is a little bottle I picked up this morning at the grocers, but I hope you will find it not too bad'; but Maurice never failed to respond with the kindest of kind words from the kindest of kind hearts, some of which he later inscribed on the fly-leaf of one of my books. It is impossible to give any adequate impression of his conversation. He was a lover of good talk, and in congenial company he flowered and blossomed. He was witty, wise, richly-allusive, generous and good. He held the connection between good wine and good talk to be close, and he expressed it in this way in the Introduction to his book on Claret:

> 'But at quite an early stage I learned one important lesson, and that is that the pleasure of wine consists only partly in itself; the good talk that is inseparable from a wine dinner is even more important than the wines that are being served. Never bring up your better bottles if you are entertaining a man who cannot talk. Keep your treasures for a night when those few who are nearest to your heart can gather round your table, free from care, with latchkeys in their pockets and no last train to catch.'

When Lord Rosebery was speaking of the oratory of Chatham he said that 'it was not merely the thing that is said but the man who says it that counts, and the character that breathes through the sentences'; and Augustine Birrell records the saying of the younger Pitt, when some extraneous person was expressing wonder at the enormous reputation of Fox: 'Ah! you have never been under the wand of the magician.'

There must be many who treasure the epigrams of Maurice, many who recall the flashes of spontaneous wit, many who recall the grace and even beauty of some of his after-dinner speeches; and many who store up in the heart the recollections of his wisdom, his tolerance and his tenderness. But the voice in which all these things were said is for ever silent; the eye which gleamed and flashed is closed; the gestures are stilled, the vehemence is no longer there; and without these things Maurice cannot be given life. This new edition of his book which contains much of his personality will recall that well-loved figure to the mind, it will bring precious and most tender memories to his many friends, and to those who did not know that brilliant figure in the glowing days of his life, it will reveal at least his universal charity and the kindliness which formed part of him.

It was entirely typical of Maurice that when he wrote about the death of a dear friend who had recently died, he said of him that 'he left many of us lonely in 1930,' and added, 'When he drank, he frequently coupled with the draught the name of an absent friend, of whom he would say some word of kindness. It was a pleasant habit, and one to be commended as a rule of practice. But the voice that proposes such a toast should be steady.' Fortunately, I am writing and not speaking, for I should find it hard to maintain the steady voice at such a time. But though Maurice has left us all lonely, his memory lives on, and this book cannot fail to perpetuate that memory. It is the memory of a most winsome and gracious personality, a great lover of life, and a great lover of his fellow men.

FOREWORD

THIS book is completed almost exactly a year later than I had originally intended. When the war came it seemed to me that this was no time to trouble a publisher with a manuscript devoted to the pleasures of the table. I threw it on one side; and then to my great surprise I found that in the eyes of some persons at least there still remained an interest in the subject, and I was urged to continue it. In the meantime other duties had claimed me, and it has only been in the past few weeks that I have found leisure to resume a pleasant task.

I owe all that I may seem to know about wine to the very kind and hospitable friends who have not been content to tell me things without showing me the truth of what they have said. The hospitality that I have enjoyed at the hands of such friends as Barry Neame, André Simon, Charles Walter Berry, Ian Campbell and a host of others is beyond all adequate recognition. Let this assure them of its constant presence in my grateful heart, and tell them further that the friendship has meant more than the hospitality. [CAMPBELL: *Of all the links which unite friendship and hospitality none is stronger or more free from petty self-interest than the common enjoyment of wine. There is no debt to pay, no boon to grant. Wine radiates goodwill to all, be they host or guest, giver or receiver: Maurice was delightful in either rôle.*]

I have made shameless use of books written by my friends. Of these I think that probably Mr. P. Morton Shand's *A Book of Wine* has been laid most frequently under contribution; Mr. Warner Allen's *The Romance of Wine* has also been of great help. Dr. Thudichum's classic has been constantly at my elbow, together with the books of Cyrus Redding and

James Sheen of nearly a century ago. Nobody would dare to write about Claret without having Cocks and Feret's *Bordeaux et ses Vins* at hand; and the writer on Burgundy must make similar use of Monsieur Camille Rodier's *Les Vins de Bourgogne*. I think that the various volumes of Constable's Wine Library are the only other printed works to which I owe acknowledgment; but I must offer my special thanks to Monsieur André Simon for help and encouragement at every stage of the work. He even read and criticized a certain proportion of the script, and would similarly have scanned the rest had time permitted me to avail of his kind offer. Such blunders as are found in the following pages are almost certainly in those he has not seen. I must also thank Mr. William Byass for some very helpful remarks about Sherry, and Squadron-Leader Guimaraens for considerable instruction about Port, which I have not altogether assimilated through no fault of his. I have a feeling that if I am literally wrong I am pragmatically right about the story of Port.

To all my readers I impart the joyous information that I have had the greatest enjoyment and pleasure in writing this book, and if they can only enjoy reading it as much as I have enjoyed writing it, we are going to have a grand time together. And their pleasure will renew and redouble mine.

God bless us all, and make us worthy of His gifts.

M. H.

September, 1940.

THE RIGHT AND WRONG OF IT

'FULCITE me floribus, stipate me malis; quia amore langueo.' So runs the fifth verse of the second chapter of the Canticle of Canticles in the Vulgate, which at Douai was translated into English, 'Stay me up with flowers, compass me about with apples; because I languish with love.' I do not know from what text those responsible for the Authorized Version were translating when they produced the familiar and sonorous 'Stay me with flagons; comfort me with apples; for I am sick of love.' The Revised Version changed 'flagons' to 'raisins,' and put a note to say that the Hebrew would be more exactly translated by 'raisin-cake'; which probably was the ground upon which some enthusiastic Pussyfoot suggested during the tyranny of Prohibition that the whole Bible should be re-written, substituting 'raisin-cake' for 'wine' wherever that blasphemous word occurred. The Jewish Encyclopædia states that the Hebrew texts differ and are in places corrupt. I dutifully accept the Vulgate and Douai; but this does not seem to me to be a text of the sort that 'Once you trip on it, entails Twenty-nine distinct damnations, one sure if the other fails.' For, even if the Catholic asks for flowers where the Protestant asks for flagons, there is one thing perfectly clear to the reader of any of the various versions of the Canticle. It portrays and commends a civilization in which the use of wine is normal and praiseworthy; it speaks of wine as an honourable gift of God; and if its author had ever suspected that there would have arisen a race of men who would consider the use of wine as unlawful and wicked he would not have known whether to laugh or cry. And let not sectarian controversy mislead us over this. If the Catholic seeks to draw a teetotal moral from the translation of verse five, let him correct himself by reading verse four. Both Protestant versions read: 'He brought me to the banqueting-house, and his banner over me was love'; but the Catholic rendering is: 'He brought me into the cellar of wine; he set in order charity in me.' So that whatever textual differences there may be between the creeds, they agree in giving us a Song in which the use of wine is praised and exalted the whole way through.

I make bold to suggest that the title of this book correctly enunciates the true religious and ethical attitude of civilized Man towards Wine. I throw down that gage to the Prohibitionists, and I warn them before they pick it up that I have this advantage over them: I am familiar with both sides of the question. I was brought up a teetotaller. [CAMPBELL: *Born, one might almost say, a teetotal crusader, Maurice, in his riper years, became a convinced wine-drinker, and the whole of this chapter on his conversion with his clear reasoning in its favour are worthy the attention of protagonists on both sides.*] In the peculiar circumstances of the country in which I was born and grew up it was right that I should have been a teetotaller. Conditions were not normal. An evil political system had produced an evil social system. For hundreds of years a people had been deliberately degraded; the laws forbade them to acquire or enjoy any substantial property; any improvement of their miserable holdings was at once noticed by the landlord's agent and the rent was promptly increased. They were forbidden all forms of culture; the priest who educated himself abroad and returned to inculcate the principles of Christianity was likely to be the victim of the first informer and to perish on the scaffold if not cut down and murdered by some drunken Hessian dragoon. Those who sneer at the power of the Irish priest with his flock ignore the fact that his influence was purchased by generations of devoted sacrifice, often sealed with the blood of martyrdom. But with all this gallantry it was little that could be done to keep alive the native culture. And wine and all fermented drinks call for culture in their use, else they will be abused. The poor wretch who lifted his heart with a tot of raw home-made Whiskey and for a moment forgot his misery was not to be blamed if he acquired a habit and took to increasing the size of the dram. Drunkenness inevitably followed; and the vile nature of the stuff consumed frequently drove the victim temporarily mad. With the reason overthrown, there was no policeman in the brain [1] to prevent crime, and crime increased. But Whiskey was not for everyone; porter, or single stout, was within the reach of all but the poorest; and as it was the only pleasure they knew, the poor creatures abused it, and drank it to excess. England in the eighteenth century knew something of this awful state; but as Macaulay pointed out, there were several degrees of poverty below the poorest Englishman before you reached the average of Irish poverty. The Englishman had other comforts and amusements besides getting drunk; the Irishman of those days had none. Fifty years ago, young men were sent off to early morning fairs and markets unfed and without the opportunity of obtaining food; and they were taught to allay the pangs beneath their belts by swilling down enormous quantities of porter, much of which usually took a return ticket. Anyone foolish enough to follow up the porter with a glass or two of public-house whiskey was courting disaster; and indeed the modern young man would do well to learn the folly of changing over from beer to whiskey in the course of a joyful evening. And it must be remembered that the Irish are a proud as well as a hospitable race. The man who accepts a drink from you expects you to accept one from him in return. Twelve young men in a public-house cannot leave until each has had twelve drinks. To go into such company without the protection of a pledge against alcohol was indeed to court disaster.

But into this dreadful atmosphere a cleansing wind had commenced to blow. A young friar named Theobald Mathew, who had been compelled to abandon his studies at Maynooth because of his giving a party in his room there, had subsequently joined the Capuchins,[2] and had become a great power amongst the poor of Cork. At that time temperance reformers

---

[1] Cf. *Macbeth*, Act I, Scene 7, lines 60 *et seq*.

[2] I hold the Capuchins in the depths of my heart, but this does not prevent me from chuckling over the libellous remark of the Abbé Coignard: 'Il faut pardonner aux capucins car ils pèchent sans malice.' *La Rôtisserie de la Reine Pédauque*.

were laughed at; but the gentle patience of that noble body, the Society of Friends, was making some little headway against the frightful abuses of strong drink which were working misery amongst the poor. William Martin, their spokesman, persuaded Father Mathew to join his temperance society, and in a short time the young priest had raised the movement into a national crusade. He took as his watchword 'Ireland sober is Ireland free'; he founded the League of the Cross, recognizing that nothing but teetotalism would save the drunkard, and that precept without example and practice would be useless. At his death he had so altered public opinion that drunkenness was beginning to be looked upon as disgraceful and teetotalism a gallant sacrifice. It was into that new world that I was born, and my baby lips were taught to repeat the pledge of Father Mathew's League of the Cross with their first lisped prayers. To the pledge so imposed upon me I was held by the circumstances in which I grew up; my father was a teetotaller, my mother drank no more than an occasional glass of wine, my cousins were all abstainers, my schoolmates were in the same position as myself. And around me was going on a very remarkable revolution; the people were gradually learning that the abuse of a thing and the use of a thing were two different things; excessive drinking was gradually declining, and the prosecutions for drunkenness were growing steadily less. With the coming of the picture-houses an alternative amusement was provided for those who had hitherto known of no place of resort other than the public-house. The war of 1914 caused some recrudescence of excessive drinking; but it quickly died down. I am told that Michael Collins imposed upon his followers the strictest abstinence; the coming of the Treaty saw the inevitable reaction against this, but under the firm hands of Kevin O'Higgins the old curse all but received its quietus. During my last visit to Cork, although it coincided with the Christmas holidays, I did not see a single person under the influence of drink. If that be a true picture of the country to-day all praise be to Father Mathew and those who so bravely followed him. But it says not one word in favour of Prohibition. A sick man will be ordered to abstain from meat for the period of his disorder; he is not expected to be a vegetarian for the rest of his life. A patient in a fever is forbidden to go out of doors; he is not to be confined to the house for ever. When a nation has been debauched into the misuse of ales and spirits it may be necessary for a time to practise total abstinence; but when the patient is better there is no longer any necessity to abstain from a gift of God. As for the pledges, there are about fifty good reasons, religious, moral, ethical and legal, why they should not bind me. The first pledge was administered when I could not understand it; the subsequent ones were all based upon the first and no free choice was offered. So long as the pledge had any value to others as an example, I remained a rigid teetotaller; it was not until I had removed to a very different jurisdiction that I allowed myself to take any form of alcoholic drink. A lawyer would probably say that there had been no consideration for my promise; if he practised on the equity side he might put it on the ground that the nature of what I was forgoing had never been explained to me. So that it will be appreciated what a quiet conscience has been filling my glass for the last twenty years.

And now I want to ask the Prohibitionists a straight question. On what principle do they justify their attempt to cut off all men from the use of something which, apart from Divine sanction, has been used by all civilized races since the beginning of recorded time? Are they content to say that evils have flowed from the use of alcohol, and anything from which evil flows should be abolished? If so, what substance will be left to us? Last year the British Medical Association informed the public in somewhat startling advertisements that bovine tuberculosis, scarlet fever, enteric fever, diphtheria, para-typhoid and septic sore throat are all borne by unsafe milk, and that about two thousand deaths a year in Great Britain alone are caused or contributed to by the use of milk. Are the Prohibitionists going to forbid us the use of milk? Mushrooms and oysters are well-known sources of food poisoning; must we deny ourselves these pleasures? And because one tin in a million of tins of food causes botulism (and I am probably putting the proportion much too high) must our larders be stripped of our prudent reserve of food? If not, I desire to hear the principle defined that permits the use of these perilous substances (and I am sure that alcohol cannot be blamed for anything like two thousand deaths a year in Great Britain), but rigidly forbids the use of alcoholic drink. If it is not to be banned upon some allegation of resultant harm, is it forbidden by any religious principle? The Mahometans have so decreed, I believe; but I am also told that various 'ways round' are found for the fervent worshippers who like their little drop. [CAMPBELL: *Readers of Alphonse Daudet will no doubt remember his story of the Arab Chief, the Aga, who, tired of domesticity and 'sick of love,' visited the Algerian* caravanserai *to watch the dancing and enjoy the wine of France. 'Cursed is a single drop of wine' in the words of the Prophet, says Daudet, but he adds that this ordinance of the Koran can be circumvented. Into each glass of wine as it was brought to him the Aga dipped his finger to the tip of which as he drew it out a drop of wine adhered; having vigorously shaken this off, he proceeded to quaff his wine without scruple.*] There is no form of Christianity that forbids the use of wine, nor was it forbidden to the Jews. If Noah was the first vintner, his drunkenness called down no curse upon the proper use of wine; for Melchisedec, the priest, chose bread and wine as the worthiest objects for sacrifice to the Most High. It is true that the ancient Rechabites vowed themselves to, amongst other things, teetotalism, much as a modern monk or nun would take a vow of celibacy, or chastity; and in each case a sacrifice was made in honour of God which for that reason was praiseworthy. But, just as every creature is not called to celibacy, so every man and woman is not called to total abstinence. Some people are natural celibates; others devote themselves to that kind of life because they feel a Call. Those who do not bind themselves to a celibate life are not thereby considered worse citizens; but they would earn censure if they were to turn their liberty into licence. In the same way, those not called to a life of abstention from alcohol are none the worse for that reason, although they would be blameworthy if they celebrated their freedom by drinking to excess. The most daring commentator on the Old Testament has never suggested that a call to abstention was addressed to others than the sons of Rechab. Nor is it their abstention from wine that the Scripture praises, but rather their obedience to the command of their father and founder. Abstention from wine was only one of their rules; they were not to build houses nor to sow seeds, nor to plant vineyards, nor have any; and they were to live in tents all their days. But mark their freedom from fanaticism. When Nabuchadonosor was advancing through the country with his army it became evident that tent-dwellers would have little chance of escaping with their lives, so they sensibly abandoned that part of their rule which bade them live in tents and joined the population of Jerusalem, where apparently they continued to reside. It recalls the priests and nuns of the period of the French Revolution who temporarily abandoned their communal life and adjusted the rules of their respective orders to the perilous days of the Terror. I have no doubt that the modern Rechabites build houses and sow seed, even if they do not plant or own vineyards; and the only part of their founder's rule to which they hold firmly is the teetotalism. Perhaps a further study of the prophet Jeremiah would open the eyes of some of their members to the tolerance of their forbears and remove a note of acerbity from their criticism of those who do not follow their practice.

The next great charge of the Prohibitionists against the use of wine is that it is not natural. In support of this allegation it is pointed out that animals dislike wine. I have never tried offering cocoa to a dog or a cat; but I have not observed any enthusiasm on the part of such animals as I have amused myself with by attempting to persuade them to drink tea or lemonade. And, if the animals are to be our judges, does this mean that we are to abandon the practice of cooking our food? For cooking is a much more unnatural process than that of fermentation; and, while I am capable of turning teetotaller, I cannot see myself sharing the cat's enthusiasm for raw mice. As Mr. Belloc so happily puts it:

> 'Birds in their little nests agree
> With Chinamen, but not with me.'

And if the animals are to be our guides in the matter of eating and drinking, are we to satisfy all our other appetites according to their somewhat promiscuous habits? Followers of this doctrine would soon find themselves on the tiles. I have known dogs that were taught to like various forms of alcohol, and I once made a Mynah drunk with a mixed Vermouth, which he wolfed greedily. [CAMPBELL: *My friend Max Graham in Oporto had a little dog called Esmé who screwed up her face in anguish whenever he offered her a taste of port, but she took it.*] I am bound to admit that his subsequent unhappy fate taught him a lesson, and nothing would persuade him to repeat the experiment. But there are a lot of things that both animals and human beings have to learn to like before they are able to enjoy them. Oily sauces repel many children; I only learned to take mayonnaise when I was about twenty. And is not wonder often expressed at the courage of the man who first ate an oyster? Many people cannot bring themselves to make that experiment, thereby depriving their lives of one of the most delightful of luxuries. But the man who will not taste Château Lafite because it is not a natural substance will clothe his body in fantastically shaped garments that interfere with his circulation and perspiration, and cut off all oxygen from the parts that most need its refreshment; he will infuse tea, coffee or cocoa by a far more artificial means than ever obtained at Château Lafite, and add milk to the result before drinking it; he will eat foods that have been prepared with so many arts and spices that no animal would touch them; he will go on his way, not on his own feet but in an electric or petrol-driven carriage; he will live from dawn to sunset shut up in an artificial prison, lit by artificial light, breathing artificial air; he will take drugs to assist his digestion, drugs to banish the resultant headache, drugs to make him sleep in this artificiality; but offer him one glass of honest Bordeaux and he will protest against this insult to nature. However, he leaves all the more for you and me.

There stands upon my mantelpiece a photograph which I purchased some years ago. It had appeared in the *Daily Telegraph*, and its beauty was such as to make me desire its permanent companionship. The lucky photographer had snapped his camera at the fortunate moment; the picture shows a young athlete who has just thrown a discus; he is poised on one foot, the other raised behind him; both arms are outstretched, and the handsome face is shown above a magnificent torso. Mercury alighting could never have looked more graceful or more beautiful. The young man was Robert Tisdall; the occasion was the Inter-University Sports, in which he captained the Cambridge team and personally won five events. When the trials were over, reporters gathered around him and asked him, as is their custom, for a message. He turned to them and said: 'Tell Lady Astor I trained on beer!' [CAMPBELL: *I have heard of a 'Varsity crew being trained on port and winning the Boat-race. Also I think I am correct in saying that A. W. ('Baby') Gore won the Lawn Tennis championship at Wimbledon on Goulet Champagne. What a tremendous smiter he was from the back of the court!*] I cite Robert Tisdall as one of the latest examples of the truth of Professor Saintsbury's assertion that 'alcoholic liquors have been used by the strongest, wisest, handsomest, and in every way best men of all times.' I confess that I was as surprised as I was delighted by his statement; for I myself should have thought that beer would be too heavy for an athlete in training, it being a very fattening form of liquor. And I should not advise any other athlete to presume that his body was made on the same pattern as the superb and statuesque body portrayed in my photograph. But it is at least permissible to base a negative argument upon the statement, and to point out that so long as the Robert Tisdalls (would that there were many such!) can drink their beer without impairing their physical efficiency, it is difficult to see how other people must inevitably damage themselves by its use.

To fanatics, arguments like this mean nothing; but then language takes on a new and fantastic meaning when used by such. I had two merry little wars in *The Listener* in 1938, the first provoked by an amazing letter from some alleged representative teetotallers, in the course of which they asserted that 'in France, where there are so many such connoisseurs, the opinion of the most eminent medical authorities is that wine-drinking is contributing rapidly to national deterioration.' *Tout ça!* as they say in *The Arabian Nights*. The present war may perhaps have seemed to encourage the writers in their belief that the French are degenerating, but the last word has not yet been said. Meantime, neither they nor anybody else, in the course of a correspondence that went on for months, my challenge being repeated every second week, have ever enlightened me as to who constituted 'the most eminent medical authorities' and when and where did they utter such an extraordinary pronouncement. It is true that the original writers, having sown their cockle, declined to reappear in the controversy; it is also true that at a later stage of it some other gentleman made reference to a congress of doctors in Poland, attended by one doctor from Lyons and two other French doctors from two places that I was unable to discover on even a very large-scale map of France; but these medicos had made no such absurd statement as that quoted, although they had subscribed to an opinion conceived in the mysterious terms that the evils resulting from excessive wine-drinking were greater than was generally believed. (I have not the quotation by me, but I do not think I have misrepresented it; the point of it lay in the vagueness of its terms.)

I stumbled by accident upon another example of the mendacity of teetotal propaganda. Without venturing to claim that I was an intimate friend of the late Lord Moynihan, I can say that I knew him fairly well; we were fellow-members of a dining club, and there and elsewhere I used to meet him about once a month. He was no teetotaller; he liked his glass of wine, and many a time have we drunk joyously together; would that those days were still with us! And Lord Moynihan once told me that even the smallest drop of alcohol would have *some* effect upon the nervous system; a man after a small glass of Sherry would not have quite the same control of his nerves as a man who had taken none. I therefore was not surprised to see in, I think, the *Daily Express* a sympathetic obituary notice of the great surgeon, in the course of which the writer said that so marked was his sense of duty towards his patients, he made it a practice to abstain *for twelve hours before an operation* from all alcohol. I am quite sure that this was true; I would point out at the same time that, as most operations take place in the morning, it only called for an abstinence from late drinks the night before. Some months later there was put into my hands in quite a different connection a copy of one of the leading teetotal weeklies. There I found this notice reprinted, under flattering headlines that proclaimed Lord Moynihan's teetotalism. I read through the reprint, and I found that they had cut out the words which I have italicized. The controversialist who stoops to such dishonesty must be badly off for legitimate arguments.

But this controversy does not any longer depend upon argument; we have had it tested, and can now speak from experience. In Norway, in Finland and in the United States of America the prohibition of alcohol was attempted. In each country after a full trial the experiment proved a failure and the law was repealed. I am told, but I do not vouch the story, that in Norway Prohibition was the result of granting votes to women, who promptly concluded that if they could prevent their husbands from spending money on drink there would be so much more for household expenses, to say nothing of finery. But the husbands, unable any longer to go out to the drink-shops, stayed at home and grew more and more annoyed with their wives; and being persons of strong views and primitive ways of expressing them, they beat their devoted spouses, first as an alternative amusement and subsequently more viciously and vengefully, until the women's vote that had imposed the law carried its repeal. There is a Biblical starkness about that story which makes me take it with a grain of salt; but I think that the underlying psychology is probably true. In Finland it was the same tale. In the United States of America more mischief was achieved by the Prohibitionists than would have been considered possible. In the days before the World War social conditions in America may have been in some parts a little unsettled; but there was a widespread regard for the rule of law, and the exceptions to this generality usually took the form of the lynching of some person whom the populace feared might escape punishment for what they believed he had done. Suddenly, silently, the Law stepped in to interfere with the citizen's liquor. He resented it. It seemed a joke. It was a law which sought to make a moral thing immoral. So he felt he was entitled to try and defeat the law. And he offered to pay good money for smuggled liquor. In my club I was asked to join in an enterprise intended to supply contraband Whiskey to Americans, and I was promised 400 per cent. on my investment. Those who did invest got seven quarterly dividends of 100 per cent. The 'bootleggers,' as they were called, became powerful and well organized. They purchased armament to protect themselves. They settled trade rivalries by pitched battle. Presently, gangster warfare having fairly come in, they took to kidnapping. Over the entire nation the law was brought into disrepute. In many states the Federal police could get no assistance from the State police. The little boy who had been taught that to break the Prohibition law was really rather a fine thing, did not exactly see why breaking any other inconvenient law should not also be a fine thing. Sensible people protested against the folly of pursuing a policy which, while bringing the law into hatred and contempt, had increased rather than decreased drunkenness. But the fanatics went on. For a dozen years or so they were permitted to sacrifice their country to their fads, and it will be fifty years before the law is restored to its proper place in the hearts of the people. The poor little Lindbergh baby might be alive to-day had it not been for Prohibition; and there were many such crimes.

The trouble arises from the incapacity of the fanatic to see facts as they really are, when he finds his own conception of them so much pleasanter to regard. He can never learn the truth of the proverb: 'You may bring a horse to water, but you cannot make him drink.' No man was ever reformed by Act of Parliament. Every really unjust law will sooner or later produce the overthrow of its author. But the Puritanical mind, which appears to be even more obstinate in America than over here, will not recognize this fact, and is, I am told, impenitently endeavouring to restore the repealed law. It is using as its argument the fact that to-day the American citizen is abusing alcohol. Whose fault is that? When it became impossible to drink legitimately, the illegitimate liquor for economic reasons had to be put up in its most compact and powerful form. Youngsters whose first acquaintance with alcohol was in the form of undiluted spirits thought unfortified wines very small beer. To meet young America in Montmartre or Montparnasse during Prohibition days was pathetic. Drunk on Pernod quite early in the evening, they would refuse all respectable forms of *boisson* on the ground that it had no kick in it. These people will have to be taught how to drink before America can hope to be truly civilized again. And the fault lies with the busybodies who took away their harmless glass of wine and drove them to more dangerous drams.

The true rule of life, in drink as in everything else, is moderation. As Pasteur said: 'Le vin est la plus saine et la plus hygiénique des boissons'; but the baby can overgorge on its mother's milk, and wine is as capable of being abused as anything else. Wine has qualities possessed by no other form of drink; it pleases at once the eye, the nose and the palate; it warms and cheers; it assists digestion; it is a stimulant and a narcotic, if I may borrow from its enemies, for it first strengthens performance and afterwards induces restful sleep. It is incapable of harbouring hurtful bacteria, and I wonder how many people who suffer from food poisoning have failed to accompany their meal by a disinfecting glass of wine. And these are but the material advantages; when we pass to the moral side, what glass can engender so much kindly charity, so many acts of benevolence, so many comforting words of friendship as the glass that has been crowned by Bordeaux or Burgundy or Champagne? So give me wine. Let my meals still be occasions of good-fellowship, where pleasant conversation will help good digestion to wait on appetite and health on both. May the gleam of a glass of Claret still attract my tired eye; may its fragrance soothe my weariness; may its savour charm my palate and carry below a spirit of concord borrowed from the feast. May I remember all the kind thoughts it inspires, for they, like the decanted wine, are free from all bitterness and have left malice behind them. And may I never drink unworthily; may I always remember that the gift which I am enjoying is not given to all; may I be prompted to help the less fortunate and to give thanks to the Giver of all good things for this, the most friendly of His gifts. Come, fill the glass. Your good health!

WINE

WHO does not remember the delight with which children slake a summer thirst with cooling drinks made of the juice of some fruit, mixed with water or soda-water and perhaps ice? A pleasure so universal must have come down to us from the earliest times; and even before the days of Noah the idea of crushing great handfuls of grapes into a drinking vessel must have occurred to our progenitors. With grapes plentiful, sooner or later some extravagant person must have crushed more than what was necessary for his immediate need; his thrifty wife probably set the surplus to keep in what corresponded to her store-cupboard. When she returned a few days later she found the juice active and bubbling; fermentation had started. The mysterious little micro-organisms had set to work upon the sugar in the grape-juice, and were turning it into alcohol and carbonic acid gas, with certain by-products. One can imagine the curiosity with which primitive man and his wife watched this eager bubbling; one cannot but admire the courage of the first person to taste the juice when the bubbling had ceased. It must have seemed odd and unfamiliar; but the sickly, sweet taste of grape-juice had gone; this new liquid was rather sour, a little prickly, and more grateful to a thirsty throat than grape-juice ever had been. And so began the practice of making wine, a process that in its essentials has not changed in the whole course of history. To-day it is carried out on a large scale, and with the aid of various labour-saving devices; there are better means of clearing the wine, and, since the coming of corks, we have learned to mature it, and thereby to improve it. But the wine that knocked Father Noah off his balance was in its essence the same thing as Château Lafite or Montrachet, its difference depending upon the grape of which it had been made and its refinement rather than upon any difference of process; for the process has remained the same. Sparkling wines are new; not until an effective method of sealing young wine in its bottle so that its self-made gas could not escape did we get the benefit of a sparkling wine. Oddly enough, this process, usually, and I think erroneously, ascribed to Dom Perignon, is attributed to another clergyman in England. Forty years ago Mr. Edward Spencer, in his amusing book *The Flowing Bowl,* told the story of Dean Nowell, a clergyman who had embraced the Reformed faith, and thus found himself the object of the attentions of Bishop Bonner when Queen Mary succeeded to the throne. The Dean had gone out fishing on the Thames one morning and had provided for the moistening of his luncheon a bottle of ale, which, fortunately, he had stoppered well, and had set in a cool place to await his desires. Suddenly came a messenger to say that the Bishop's minions were after him; the Dean, obsessed by thoughts very remote from ale, fled the country and remained away until the reign of Queen Elizabeth gave him an opportunity to return. Resuming his old habits, he went fishing on the Thames once more, where he was wont to tell the story of his escape; and one morning, by way of adding a dramatic touch to the story, 'and here,' said he, 'is where I had set my dinner'; and there was the bottle of beer! But, as soon as he sought to open it, in order to drink a grateful health unto Her Majesty, the bottle exploded in his hand, without doing him harm, however, and leaving enough of its contents for him to taste bottled beer for the first time. It is a pleasant, and not improbable, tale. Shut up an incompletely fermented liquor in an all but airtight container and it will continue to turn its sugar into alcohol and carbonic acid gas, which may even burst the bottle or barrel, if generated in sufficient quantity. So long as it was the custom to stopper wine bottles with a drop of oil and a morsel of tow, the gas would escape as fast as made. As soon as corks came to provide an airtight seal the phenomenon of a sparkling wine followed as a matter of course. Dom Perignon rightly owes his fame to the fact that he was a blender of uncanny skill, and knew the best proportions in which to mix his grapes before setting them to ferment. But there may perhaps be found in his skill as a blender the explanation of the undoubted fact that Champagne can deal with its own gas in a way that has been achieved by no other wine. I remember a delightful holiday spent in a pre-Hitler Germany, when my companions patiently sought for a German Champagne that would satisfy them. We must have tried a dozen different kinds; and on other occasions I have given careful attention to sparkling Mosels and Muscatels, wines from Touraine and elsewhere, all vainly seeking to compete with the wines of Rheims and Epernay. I found them mawkish and sickly, and they were almost always productive of a headache. But real Champagne may be drunk with impunity in any conditions; and, in the absence of excess, it will refresh and cheer without giving rise to any subsequent unpleasantness.

There is one feature about the making of Champagne which should be noted. Champagne is a blended wine, each shipper having his own ideas as to the proportions of this and that grape; so that he will draw his supplies from many and distant parts of the Champenois country. But the grapes cannot be got to the press too quickly, because if they were allowed to remain in their baskets until they reached Rheims or Epernay, fermentation might have set in, and then the colour of the skins of the red grapes would have indelibly coloured the wine, whereas the pinkish tint caused by the pressing of the red grapes is removed by the subsequent fermentation. When I read of imitation Champagnes made in this country out of dehydrated French grape-juice I always wonder how this difficulty is overcome, and whether it is this that makes the English product taste so unlike genuine Champagne. In fairness, however, I should mention that I have been told by no less a person than Monsieur André Simon himself of a new American process for preserving vegetables, which consists of extracting all the water at the time of gathering, and restoring the water by soaking just before preparing them for table. He tells me that the invention is so successful as to deceive the elect. He did *not* tell me that grapes could be treated in this fashion, nor does it seem to me a likely thing. However, we live and learn; I am old-fashioned, and I am content with wine made in the old-fashioned way.

Wines are classified in different ways: red and white, still and sparkling, natural and fortified. All wine would be white if the skins of the red grapes were not allowed to ferment with the fruit; the Douai version of the well-known verse: 'Look not upon the wine when it is red' is 'Look not upon the wine when it is yellow'; and I think that all Biblical wine was what we call white, for it was the custom of primitive peoples to press their grapes, and our red wines owe their colour to the custom of allowing them to ferment without pressing the grapes, the skins of which dye the wines in the process of fermentation. [CAMPBELL: *This is too sweeping a statement. Vineyard proprietors, like other wise men, are no doubt constantly trying out suggestions for the improvement of wine-making, but I hope the day is far off in which the grapes which produce my port wine, for instance, my rich red port, are not pressed in the* lagar *by the foot of man, the most natural, the best and the surest of all methods.*] There are three black grapes to one white in Champagne. The difference between still and sparkling wines arises from the stage at which the wine has been bottled. If it has been allowed to turn practically all its sugar into alcohol while in the wood the gas will all have escaped, and the wine that goes into the bottle converts the remainder of its sugar in other ways than fermentation. If, on the other hand, the wine has been bottled when only the first wild fermentation has taken place a great deal of unfermented sugar will be present, and this will continue to convert itself into alcohol and gas to such an extent as to blow out the cork, unless this has been wired down. Any wine can be turned into a sparkling wine by bottling it at this early stage. But some wines are not suitable for this treatment, and only Champagne can undergo it to the general satisfaction.

The difference between natural and fortified wines is profounder. Claret, Burgundy, Champagne, Hock and Mosel are all natural wines, because nothing is added to the juice of the grape in the course of manufacture. This list is not exhaustive; indeed, it might be fairer to begin at the other side and say that Port, Sherry, Madeira and Marsala are fortified wines, because a spirit distilled from the wine is added at one stage or another of the process of manufacture, which adds to the alcoholic strength of the wine, and gives it a distinctive characteristic. There are purists who deny to Port the right to call itself a wine, because the Brandy is added to the must before the latter has made any progress in its fermentation; Brandy, distilled from Consumo wine, is added to the must on the second or third day of the fermentation, with the result that fermentation practically ceases, and the wine goes through a long struggle between the Brandy and the sugar which is trying to ferment; and this battle may last for ninety years or more. For this reason, nobody drinks a vintage Port until it is at least about twenty years old; and it will probably not reach its best until it is forty. In Sherry the Brandy (made in this case from young local wines) is not added until the fermentation is complete; strength is gained, but sweetness is not preserved. Sherry, however, is of its nature a very long-lasting wine, and, even without the added spirit, would probably attain a great age. It takes an expert to tell the difference between a venerable Sherry and an old Madeira; both are remarkable for marrying the qualities of richness and dryness. Marsala is the Cinderella of fortified wines; I often wonder that some merchant does not specialize in it, for it can be excellent, and grade for grade it compares very favourably in price with other fortified wines. I can remember an Old Particular that was the equal of a very fine Madeira, yet it only cost ninepence a glass.

Now, what process in the world could be more natural than that of the making of an unfortified wine? The grapes receive neither more nor less cultivation than those served to the teetotaller for his dessert; indeed, if there be a difference, the dessert grapes may have been more elaborately manured, for nothing but the plainest of road-sweepings helps the wine grape along. In its picking it receives about the same degree of care; and, before it is left for fermentation it undergoes no other treatment than that of being passed through a device which tears away the stalks and thereby slightly crushes and opens up the berries. Formerly it was an almost universal practice to crush the grapes by treading them underfoot in stone troughs. Gradually this custom is dying out; it is unusual to-day in the French wine districts, although it still continues in Spain and Italy. In France, in the case of white wines (and this is also true of Hocks and Mosels) the grapes, once free from their stalks, are put through the wine-press, and the must is left to ferment in large casks, where it gurgles and bubbles for a fortnight, by which time the first fierce fermentation is over. All that remains to be done to a still wine is to allow it to rest in cask for a couple of years, occasionally drawing off the wine from the lees into a clean cask, and once or twice 'fining' it by stirring in some isinglass or similar substance, which forms a sort of net in the wine and draws down with it most of the finer particles of the *débris* caused by the chemical changes. Then the wine goes into bottle, where it rests for a few years further before it is drunk. With the exception of the fining, Nature has done the whole thing. There would seem to be no room for an Escoffier or a Latry in the preparation of wine. Nevertheless, wine can be made badly and it can be made unusually well; skill is never without its reward. In the chronicles of Bordeaux there are some names so recurrent as to suggest hereditary skill.

Wine in its turn can be turned into Brandy (Brent Wein: Burned Wine) by the process of distillation. I do not recommend the practice as a hobby or a home industry; apart from legal difficulties, the home product would taste uncommonly nasty. But if you possess a chemical retort and put wine in it and set a spirit lamp underneath you can draw off the spirit of the wine, which, when suitably purified, is Brandy. Brandy may be made from any wine; but only the Brandy distilled from the harsh and unpleasant wine of the Charente is entitled to the name of Cognac, which is a place-name that may not properly be applied to the Brandy of any other district. Cognac is the best Brandy in the world; Armagnac, made down nearer the Pyrenees, is its runner-up, and a first-class Armagnac is better than a third-class Cognac. They make a very palatable Brandy in Spain, some brands of which are popular in England. The taste is very distinctive, and the Spaniards would be foolish to attempt to sell their product under a Cognac label. [CAMPBELL: *Some Spanish Brandy houses used to sell it as* Coñac, *but that is, I believe, now a thing of the past.*] Being older and wiser than the Australians, they never do so. In the Champagne and Burgundy districts a kind of Brandy is distilled from the murk left in the winepress when all the juice of the grapes has been extracted. Water is poured on the pulp, sugar added, and the mixture is distilled. The product is called 'marc,' and I do not think I have ever met a marc that was not described as a 'vieux' marc. The liqueur is very popular in France, but, although I have tasted marcs that were better than others, I cannot say that I really like it. To my palate it always tastes as though it had been distilled from laurel leaves. Marcs are made in many wine districts, but, with the exception of one Italian one, that tasted like an Abyssinian atrocity, I have never come across them outside of France.

Distillation can be applied to many substances, some good and some not so good. Whiskey stands to beer in the same relation as Brandy stands to wine. But Whiskey rarely attains the delicacy appropriate to a liqueur, although I have many times deceived the elect by serving an Irish Whiskey after dinner. On the only occasion when this trick was attempted to be played on me I detected it, and I confess I see no great matter for boasting in this. The flavour seems to me to be entirely 'other.' But my own guests who fell included some accepted authorities. The first time I played the trick it was with a bottle of John Jameson that had been distilled in 1893, which was given to me by the late Francis Berry, God rest him, in 1926. The second time I tried it with a bottle of twenty-one-year-old John Jameson that my uncle had brought me from the Vice-Regal Lodge, although, oddly enough, he had re-imported it from Wisbech. Grown bold, I tried a third time with the ten-year-old John Jameson that is procurable in most places. Once again the deception succeeded and I was complimented on my Cognac. But as this was the liquor with which the late Dominic Daly of Cork had deceived one of the Hennessy family into believing that he was tasting a good Cognac of the 1893 vintage, the result is not a matter of marvel. [CAMPBELL: *This mistake is always being made. Our regimental Medical Officer in the trenches in* 1915 *had the Commanding Officer and others into his dug-out to inspect his outfit. With the exception of myself they were all habitual whisky drinkers, but they all thought (and so did I!) that the whisky shown us was brandy and passed favourable judgment on it as such.*] I do not drink Whiskey myself, but I keep both Irish and Scotch; and I find that those of my friends who are persuaded to try the Irish rarely return to the Scotch. Irish Whiskey is made in the old pot-still, which cannot extract spirit from any but genuine materials. A patent still can do a lot with potatoes and even wood. I doubt that anybody would mistake any but a rare 'single' Scotch Whiskey for a Cognac. By a 'single' is meant an unblended Whiskey.

I have been endeavouring to show that alcoholic beverages, honestly made, are natural; this is especially true of wine, and their distilled products have had little more attention from the art of man than has an egg when it is boiled. But I hope that nobody has so misread me as to think that I am recommending the partaking of these gifts with the carelessness with which a man might sit down and eat a dozen boiled eggs (don't start: I have often done that). The advice that I frequently give to young men is: don't begin drinking at all until you are seventeen or eighteen, for you are not old enough to appreciate what will be given to you. Drink only wine, beer and cider until you are thirty, and keep the reinforcement of spirits for a time when your system may be glad of such assistance. And even in your first stage never drink for the mere sake of drinking, and NEVER drink with the intention of getting tipsy. A glass or two of wine will make the heart beat a little faster and will promote a sense of good humour and jollity. But to think that because two glasses produce ten degrees of gaiety four glasses will produce twenty degrees is a great delusion. People vary as to the ration of wine that is good for them, and as soon as they take more than is good for them the effect is bad, as it is if they take too much beef or bread or tea or milk. Therefore, it is important to learn as early as possible what one's proper ration is; learn it, and do not exceed it. But mark: the ration appropriate to a solitary meal will be smaller than that suitable at a party; I fancy that the gaiety of a party has an effect upon the system and makes it capable of dealing with larger quantities. I know that I should very rarely feel inclined to drink more than one half-bottle of Claret if I were dining alone; but if there were two of us I would always decant at least a bottle and a half, and possibly two bottles, and I would certainly have two bottles amongst three, and, I think, three bottles amongst four. Another peculiarity about one's appropriate ration is that it alters according to latitude. My half-bottle in London becomes a bottle in Paris; and I should not care to deny that it might double itself again in Bordeaux.

Another rule I try to teach my young friends is to eschew the cocktail before drinking wine. I am no confirmed enemy of the cocktail; there are occasions when it not only pleases but also fulfils the useful purpose of giving a tired man a fillip before he meets his guests. [CAMPBELL: *No, Maurice, if you wish to encourage the young to drink wine you must definitely discourage the drinking of cocktails: it soon becomes a habit difficult to eradicate.*] But even then I would choose something like a 'Sidecar,' made of equal quantities of Brandy, Cointreau and lemon-juice; because the spirit is a wine spirit, and will not quarrel with the wine as Gin or Whiskey would. A Champagne cocktail is still less objectionable; only the bitters call for a rebuke. But all cocktails with spirits in them are subject to this criticism: they attack the walls of the digestive tract too violently, and partially paralyse the functions of that tract. If you have a tender spot on your skin, a scratch, for example, try the difference of spilling a little wine on it and spilling a little Brandy. Note the sharp sting of the spirit. Now try this experiment. Take one of each of two different kinds of chocolate from a mixed box, taste them, and observe the marked contrast in their tastes. Then take a sip of Brandy, roll it around your mouth, and immediately take another pair of chocolates. You will find that the difference between them is much less marked. The taste nerves are drunk; they are not doing their job. For which reason it is madness to take spirits before a meal at which delicate wine is going to be served. To a lesser degree the same criticism may be levelled at any very strong-flavoured preface to the wine; the flavour will not paralyse the taste nerves, but it will tire them; and these strong flavours have a habit of revisiting the palate at a moment when their presence is most undesirable.

Another tip to the neophyte is, don't waste your good wines. If you are going to drink wine every day, and in this country beer is a more natural daily beverage, let your daily wine be an *ordinaire;* most wine merchants ship a sound Claret or white wine at about half a crown a bottle or one-

and-six a half-bottle that should be quite good enough for daily requirements. [CAMPBELL: *Not to-day, alas; the price of* ordinaire *Claret to-day is some four or five times half-a-crown a bottle. And better wines in proportion.*] But when a friend is coming in have something better for him. No need to give him Montrachet or Château Cheval Blanc 1921; for the ordinary dinner young men exchange the wine need not and should not cost more than 4s. a bottle. For the really great occasions bring out the bottles from behind the faggots. On these be lavish; you are trying to celebrate a notable occasion, and don't spoil the ship for the want of a ha'porth of tar. But even here a half-sovereign should get you something sufficiently outstanding. I mention these prices because if you look through the catalogues of several wine merchants you will find offered some rare and costly bottles for which you will be asked to pay anything up to three or even four hundred shillings a dozen. I do not think that at the moment of writing I could buy Château Cheval Blanc 1921 for less than 30s. a bottle, or Château Yquem 1921 for less than £1. Yet I know of one cellar in London that stables hundreds of these White Horses and also shelters hundreds of the Yquem. But the owner is a very rich man, who is expected to offer gold and myrrh and frankincense to his guests; and if he produced mere ten-shilling bottles his commercial credit would probably suffer! It is right for such people to offer their guests treasures that they can afford; but it is both foolish and unnecessary for other people to do so. Incidentally, it is in the pages devoted to Hocks that you will find the most fantastic prices quoted; and no German wine is worth such prices.

On occasions of elaboration, be careful to serve your wine against an appropriate food; but when dining alone or intimately do not take too much trouble. The theory that you cannot enjoy a Claret with fish is perfectly sound when you are drinking something like Château Ausone 1899, which Monsieur Boulestin once found his guests attempting to drink against sardines! But I never find anything to regret when I pour a little bourgeois Claret to accompany a bit of sole, or even salmon. Few strongly flavoured foods make a good background for red wine, and sweets of any sort destroy it.

On the other hand, cheese and mushrooms make so good a foundation that they will even redeem an accompaniment otherwise inappropriate. Eggs are undesirable on account of their sulphur; but an omelette made with cheese or mushrooms will help a Claret right cheerfully. White wines are less exclusive in their companionship, although I never enjoy them with meat or even chicken. And I never know how best to serve the greater Hocks (when I am lucky enough to possess any!) for they seem to me to be wasted on the fish, not virile enough for the entrée or roast, and too austere for the sweets or dessert. They fall between two stools; Montrachet or a good Chablis or even a Meursault defeats them in the lists of dryness, while in a contest of sweet wines Yquem and its supporting Sauternes canter easily home ahead of a breathless Schloss Johannisberg or Ruppertsberg. Once a Hock stood me in good stead. I was showing off a cook before a good friend of mine who was a gourmet and about to pension off his own chef. My candidate prepared a menu calculated to exhibit his own skill, but taking no forethought as to how my cellar could support it. The Clarets had finished with an 1879 Château Latour, drunk with a saddle of mutton; there followed a 'Halali' salad, a delightful compound of the white meat of chicken, mushrooms, pineapple and other things, rather rich and sweet, yet not quite a sweet. This was to be followed by a Pêche Melba, with which we were to drink a bottle of Yquem 1921. The problem was what to serve with the salad. A bottle of Ruppertsberger Hoheberg (Riesling Auslese) 1921 did the trick to perfection; sweet enough to match the pineapple in the salad, it led up to the sweeter Sauternes with a self-effacing chivalry that one does not often associate with its race. It is one of my pleasantest memories of the Rhine wines.

There are simple rules, made, like all rules, to be broken in the proper circumstances, which govern the serving of wine. Sherry is the best wine with which to open a meal; a very dry glass in the ante-room can be succeeded by one less dry with the soup. Madeira competes with Sherry as an *apéritif*, and also as a soup-squire, but, unless the Madeira is unusually good, in which case its proper place would be at dessert, I prefer Sherry. These wines may be drunk with oysters, but I prefer a still unfortified white wine—a Chablis or a good Pouilly, or a Champagne, although I think the sparkle does not suit the oyster: it irritates his last moments when he is sacrificing himself for our enjoyment and deserves our consideration. By the way, I ought to have mentioned Chambéry as an agreeable *apéritif*, which is not as widely known as it might be.

With the removal of the soup the serious business begins. If it is a Champagne dinner, that wine may be served right through the meal; but for Heaven's sake do not follow it with Port. [CAMPBELL: *This is a matter of individual taste and there is no 'thou shalt not.' I enjoy my vintage port after my champagne, particularly if they are separated by a good claret.*] No two wines disagree more certainly or persistently than these; and yet it was Soult, and not Lannes, the gallant Duc de Montebello, that carried war into the country around Oporto. If Champagne is not to be the wine of the evening, a glass of it with the fish will not be out of place, and it will set tongues wagging more speedily than any other wine; but something less plutocratic will match the fish equally well and less ostentatiously. My preference is for white Burgundies or Rhône wines, or something from the Loire, or even a good white Rioja (I have drunk excellent Riojas, procured at very little cost). Certainly not a Sauternes or Barsac, though there is a school of thought that supports their use; and you may give the Graves to my neighbour. [CAMPBELL: *This is also a matter of taste and unfair to Bordeaux white wines which go well with most fish. Alsatian wines might also have been mentioned in this connection, but had not perhaps established their place in Maurice's day.*]

The entrée will introduce your red wines; Claret before Burgundy, and a light wine before a fuller one, except that youth precedes age, a delicate old Claret setting a crown on any feast. Give your white wines half an hour in the refrigerator before serving them; bring your red wines up to the room in which they are to be drunk (or even one slightly warmer, like the cooler corners of the kitchen) about twenty-four hours before decanting them. Never warm them, either by setting them in front of the fire or by giving them a warm bath. Wine is a live thing, and sudden changes of temperature are very harmful to its flavour and bouquet. Sherry, being a white wine, comes within the icing rule; it should be served very cool; but Madeira appeals to me more if served as a red wine, *chambré*, no more. [CAMPBELL: *If* chambré *simply means at the temperature of the room I might agree, but I would rather have it a little cold than a little warm.*]

Port is an English tradition. Port should never be served at a dinner where Sauternes is offered with the dessert; the wines kill one another. Port drinks at its best when October has settled in, and the only merit I have ever found in an east wind is the marvellous effect it has in making this wine more grateful and comforting. Port enjoys in this country the reverence more properly due to Claret; for in praising a fine bottle of Claret you are giving glory to God, but Port, under God, owes much of its excellence to the skill of its shipper. But I should be sorry to discourage the reverence accorded to Port in places where Tradition still holds sway. Many an honest drinker, bred upon Port, has felt his way towards Claret, and has died in the true faith. So drink your Port without smoking; eat a Cox's Orange Pippin or any good, firm, sweet apple with it, or crack a few nuts, if your dentist still maintains your own teeth for you, and has not fitted you with a plate to catch every pip and stone and fragment of nut that can be stowed in so limited a space.

Whether the wines conclude with Port or Sauternes or an old Claret or Burgundy, follow them with nothing but good vintage Brandy.

SHERRY

OF all wines, Sherry is the maid-of-all-work. There is no better expression of casual hospitality than a glass of Sherry; and on the other hand the wine can appear without impropriety at almost any stage of a dinner, although more usually limited to the beginning and end of the meal. [CAMPBELL: *A light, rather dry, Fino sherry is good drinking all through a meal. I have tried and enjoyed it.*] But Professor Saintsbury has designed quite a respectable course of Sherries to carry the diner right through the feast; and I am prepared some day to try his experiment. The reason for this versatility is to be found in the almost unlimited variety of form and flavour to be found in this one wine. You can find a Sherry so dry as almost to rasp your palate, and so light as to conceal its real alcoholic strength; or you can go to the other end and drink a rich and sweet wine that will compete with the heaviest vintage Port. In between are all the varying hues of the rainbow. And if ever you find a Sherry that specially pleases your palate you can take a sample to your wine merchant, and, even if you are unable to give him any particulars of its pedigree or provenance, he will be able to procure for you an almost exact match. For Sherry is a fortified, blended wine, which owes its individuality to the skill and taste of the shippers who prepared it; and, in almost every case nowadays, it is not a wine of a single vintage year, but a judicious composition into which the succeeding years have each cast their blessing.

It is obvious, therefore, that Sherry is a wine of a nature entirely different to that of Claret or Burgundy or the German wines; different with another difference from Champagne, which is also a blended wine. To say 'I like Sherry, but I don't care for Claret' would be like saying, 'I like coffee, but I don't care for lager.' It is hardly correct even to compare Sherry with Port, another fortified wine; for while Port takes its dose of Brandy while it is still must that has hardly started to ferment, Sherry is allowed to complete the first part of its fermentation before the spirit is added. If no Brandy were added to Sherry it would probably show a remarkable affinity with the wines of the Jura, upon which Pasteur carried out his famous experiments. Those wines, without any added spirit, attain unbelievable ages; sixty and seventy years is quite usual. It can readily be imagined that Sherry, with its fortification, can go to ages that are far greater; centenarian Sherries are only uncommon because of our own impatience and the laws of economics.

Sherry is a corruption of Xeres, which the Spaniards, regardless of good English (witness their obstinacy in speaking of Trafálgar as Trafalgár), insist on pronouncing Heréth. From a not-far-distant area, of which the port is San Lucar, come the wines called Manzanilla, to which the purist denies the name of Sherry, although the affinity is very close. The Manzanilla wines are extremely dry, so dry as to be almost bitter, and I am told that their distinctive flavour resembles that of camomile, a drug that I have had the good fortune to avoid, although, if allowed to take it in the form of Manzanilla, I should not shy at the prescription. Inland near Cordova is a district called Montilla; its wines awakened the competitive spirit in Xeres, where the shippers produced a wine to match the Montilla wines, which they christened Amontillado. To-day I think Montilla has become a description of type and flavour rather than of place of origin. Although usually deserving of the epithets 'light and dry,' I have generally found it applied in this country to a slightly fuller wine than that which is labelled Amontillado. The pupil has gone beyond the master.

Sherry, however, is not to be satisfactorily discussed in terms of place-names, for, as I have said before, the wine we know to-day is a blended wine, and I fancy that the various districts may all be laid under contribution for any one well-known named Sherry. For Sherry is made upon what is known as the Solera system; and in passing it may be no harm to warn the neophyte that a Solera does not mean a type of Sherry, such as Amontillado, or Montilla, or Oloroso, or Manzanilla; there are Soleras of each of these kinds of Sherry, and it is almost impossible to get a Sherry to-day that is not a Solera. Mr. Warner Allen has hit it off admirably when he says in his volume on 'Sherry' in Constable's Wine Library: 'The various wines hitherto described are, as it were, the alphabet which the shipper employs to produce the Soleras that may be compared to words, while the finished blends composed from the Soleras are his sentences.' The only qualification I should like to put upon that excellent simile is, that sometimes the wine sold by your wine merchant may be described as a 'Solera'; probably it will be called 'Old Solera' or 'Solera 1875' or some other year. If that description is accurate, it simply means that a particular Solera has itself attained sufficient charm and individuality to be shipped without further blending. How the Solera is formed we shall see in a moment.

But before we come to the making of Sherry, let us first cast a glance at its history, and, if one may use the word, its ingredients. Mouillefert gives a list of no less than twenty-two different grapes that go to the making of wine in different parts of Spain. Their cultivation goes back to pre-Roman days; but Pliny, who writes of the wines of Laletani (which I believe was Barcelona) and of Tarragona, has nothing to say about what is now the Sherry district, although it is safe to assume that the vine was already under cultivation in those parts. I was somewhat surprised to learn that the Moors, during their occupation, did much to further and improve the wine trade; Mouillefert says that in his day, a couple of generations ago, the remains of the Moorish vineyards were still shown to visitors to Malaga, although they had been almost entirely destroyed by the phylloxera.

The vineyards of the Sherry district are divided into three classes; the best, called Albariza, have a chalky soil, while the second best, called Barros, are principally clay. The poorest, called Arenas, have a sandy soil. Although a variety of vines are planted, the informing grape is the white Palomino; unlike most other wine districts, the Sherry vineyards are planted almost exclusively with white grapes. But the fruit is not rushed from the vines to the press, as in

Champagne, where they dread the possibility of fermentation beginning in the basket; our Spanish grapes are carefully collected and laid out on straw mats to be sun-dried after being detached from their stalks. One would expect this extra sunning to produce a preponderance of sugar; it is therefore strange to read in Thudichum that, unless the shipper has deliberately added sweet wine to his blend, Sherry wine is never sweet. This may be the origin of the word 'sack' (*secco*), for when an Englishman, accustomed to his gentle Bordeaux, suddenly met the fierce embrace of Manzanilla, he must indeed have been tempted to describe its flavour in words wherein 'sack' would probably have sounded mild. I think that there is very little doubt that the Elizabethans used to sugar their Sherry; certainly the impression conveyed by the phrase, 'a cup of sack,' is not one of dryness. In passing, it will of course be appreciated that Shakespeare was guilty of a shocking anachronism when he made Falstaff drink sack; the Spanish wines were not sold in England in Shakespeare's day. If some modern author, writing about the Duke of Marlborough's campaigns were to describe him as telephoning to one of his subordinate generals it would hardly be a greater anachronism. But it is partly because of their anachronisms that we love the Elizabethans.

When the grapes have been sunned enough they are gathered up and emptied into wooden 'lagars' about ten feet square. The lagars have sloping sides about two feet deep, and in the centre arises a large vertical screw, with a sort of glorified wing-nut on it, this screw being twisted up to its full height to be out of the way of the vintagers. As the grapes lie in the lagars they are covered with a couple of shovelfuls of 'yeso,' or burnt local earth, which is very chalky, and is, I suppose, an antacid. The practice is as old as antiquity, and is agreed to be entirely beneficial; the sulphate of lime acts upon the must and counteracts any excess of acidity. When the yeso has been allowed an opportunity to work itself in, the vintagers climb into the lagars, and then begin a treading of the grape that is quite unlike that with which the poets have made us familiar. For 'the stray feet' that 'pulp the globed weight of juiced Iberia's grape' are not the bare feet of laughing boys and girls, as the poet would have us believe; the vintagers of the south wear shoes, heavily studded with nails, and only when so clad do they perform their ritual trampling of the grapes, which quickly yield their burden of juice. When the vintagers find that they are treading murk they rake the stuff together and build a careful column of it around the screw, which they bind tightly with esparto grass; they then turn the screw, until the last drops have been squeezed out through the puttee of esparto grass. If the murk has been badly built up, or if the puttee has been unskilfully wound (what memories can a phrase recall!) the pillar will collapse, and all will be to do again.

Then begins the most fascinating of all the stages in the manufacture of Sherry. If a Bordeaux vigneron gathers his Cabernet Sauvignor grapes and sets them to ferment he may not know whether he is going to get a good Claret, but at least he knows he is going to get Claret, and he would be rightly dumbfounded if the wine proved to be a Sauternes. Yet I think that there is as much difference between 'Tio Pepe' and 'Bristol Cream' as there is between Château Lafite and Château Yquem; and nevertheless, when the Sherry shipper sets his must to ferment he has no idea whether he is going to get a sweet wine or a dry wine. There are three types in the Sherry classification at vintage time; the wine may prove to be Fino, or dry and light, Palo Cortado, or medium, or Raya or rich. And no living being can tell you what a particular vat of must is going to turn out to be; and if there have been filled at the same time with the same must three vats, one may prove to be a Fino, one a Palo Cortado, and one a Raya. It is one of the mysteries of nature for which no explanation has yet been found. To the watcher, the vat seems to bubble merrily for a few days; when the turmoil is settling down a little Brandy (distilled from the local wine) is added. Soon there appears upon the surface a horrible-looking greasy white scum, which is called the 'flor'; this is encouraged by departing from the practice obtaining in other wine districts, where the air is kept from the wine as much as possible; here it is given free access. The appearance of the 'flor' is a moment of great anxiety, for there are tiny organisms called the *mycoderma aceti* and the *mycoderma vini*, either of which may appear at this stage, and the former will turn the vat's contents into vinegar, the latter into wine. The owners watch this repellent scum, that looks like a mass of baby oysters floating in grease, with the greatest anxiety. At length it settles down, sinking to the bottom of the vat; the contents are now racked off, and for the first time the shipper discovers to what type the wine conforms. The Finos, Palos Cortados and Rayas are segregated, and what in any other district would be called the making of the wine is completed.

But so far as Sherry is concerned, the labour has only begun. The artist has mixed his colours and prepared his palette, that is all. These young wines go to form the cradles, from which are drawn the requisite supplies that go to making the Soleras. Now, how is a Solera made? Well, to begin with, you must grasp the idea that a Sherry of a particular Solera never reaches the customer until it has passed through about a dozen successive blending vats. Messieurs X, the well-known shippers, may decide this year to begin a new Solera that they will put on the market in ten or twenty years' time. They experiment, taking two jars of this wine and one of that, adding a little *arrope*, which is wine that has been reduced to about two-thirds of its bulk by boiling, being thus very much strengthened in flavour, and perhaps adding some of the very sweet wine made from the *Pedro Ximenes* grape: in short, playing about with the ingredients until they get a flavour that they think the public will like. Then they fill a vast *tonneau* with this mixture and leave it to mature. From time to time they will build up by its side successive vats full of the same blend, but as time goes on the younger wines will be taking the place of their predecessors. The blend, it must be observed, is composed not merely of wines of different types, but also of wines of different ages, quality being improved by the proportion of old matured wines that is used. When the young wines are sent to the cradles they are 'fixed' by the addition of more Brandy, so that the ingredients are more constant in quality than would be possible or probable in an unfortified wine. As a result, in ten years' time there will be ranged side by side perhaps a dozen large vats each full of a blend of Sherries made according to the same formula, and only differing by their decreasing age. When at last the Solera is put upon the market all the wine offered to the public will come from the oldest vat, which will be filled up from the next oldest, and that in turn from the next, and so on, the youngest vat being filled up from the cradles, on the basis of the original recipe. Now, when a blend of wines has passed gradually and successively through twelve different containers, being subjected to a new mixing by reason of each transference, it is easy to see why the wine drawn from the twelfth vat should be constant in quality and flavour. That is what makes Sherry unique amongst wines, and explains why one can continue to ask for a particular mark with the certainty that there will never be an appreciable difference between one bottle and another.

Sherries so blended are shipped in wood, and bottled as required for sale. The next time you see a bottle of 'Tio Pepe,' look closely at the label; you will see a combination of letters perforated in it. By this the age of the wine in bottle can be ascertained by those whose business it is to know; for Sherry should be drunk shortly after going into bottle, or should be kept in bottle long enough to develop bottle age. In the intermediate stage it is not at its best. It can be kept for three or four months without hurt; but the wise course is to order one's Sherry in comparatively small quantities, not more than a few weeks' supply at a time. The wine does not need a rest on arrival, for the occasion for that rest, so necessary with certain other wines, is to enable a disturbed deposit to settle down again. In the case of Sherry, even a very old wine throws little or no deposit. When sugar is dividing itself into alcohol and carbonic acid gas the chemical process causes a certain amount of what may well be called dust to settle in the wine. In Claret or Burgundy, which are unfortified wines, an appreciable amount of sugar remains unconverted when the wine is bottled, and continues a very slow process of fermentation in the bottle, which causes a considerable fine sediment to form and in some cases a thin crust, this being contributed by the fine *débris* of the grape skins. In the case of Port, an enormous proportion of sugar remains unfermented when the wine is bottled, because Brandy has been added to check the fermentation at a very early stage. The result is that Port, continuing its fermentation in the bottle, fighting the added Brandy, throws quite a heavy crust, and forms a substantial sediment as well. Hence the religious reverence with which a bottle of Port must be handled—a reverence that one sometimes sees mistakenly offered to a bottle of wood Port, which needs no more careful handling than a bottle of Sherry. [CAMPBELL: *Personally I consider that all wine, sherry and wood port included, should always be carefully handled.*] Sherry, having completed the transubstantiation of almost all its sugar before being finally dosed with Brandy, and having that Brandy to make still slower the conversion of the rest in bottle, throws practically no deposit at all. On the one or two occasions when I have had the good fortune to come across very old-bottled Sherry I found that the last dessertspoonful or so was slimy and a little thick; but the rest of the bottle ran free and clear.

Old-bottled Sherry is a thing apart, most rare and difficult to get. I do not refer to the wines so described in the average catalogue, for these rarely pretend to a bottle-age of more than twenty years, and a great deal more than that is necessary to endow the wine with the quality I mean by bottle-age. I have affectionate memories of two bottles. One was given to me by a kind fellow-guest at a dinner, who learned with appreciative interest of my affection and reverence for wine. He was himself in the seventies, and he told me that this wine had been in his father's cellar and his own as long as he could remember, and the first thing he remembered about it was that it had been described to him as very old indeed. I have no reason to doubt that when I drank it I was drinking a wine of more than centenarian age. Now, can you imagine yourself drinking liquid honey—but a honey that had lost all its sugary flavour, and, while remaining rich, was perfectly dry? A more indescribable flavour I have never met; but it was entirely delightful, and with every sip, and there were many, I blessed the kind donor of such a source of innocent delight. It had lost nothing of its strength; indeed, if I had been told that it had gained in alcoholic content I should not have been surprised.

The other bottle took its place in a procession of great wines offered to a party of his friends by Doctor G. C. Williamson at his delightful home, Mount Manor House, Guildford, on September 19th, 1935. It was described as an Amontillado, 1847; this probably meant that the Solera had been formed in 1847, although at that period Sherry may still have been shipped as a vintage wine. The important thing was that it had come from the Duke of Devonshire's cellar, where it had lain in bottle for a great many years before our host had bought it another long time since. Its bottle-age must have been at least fifty years. Writing of it at the time in the *Wine and Food Quarterly*, I find that I described it as 'full and luscious, yet absolutely dry,' which would equally have applied to the other bottle of which I have spoken. There is a majesty about these venerable Sherries that is peculiar to them; I know of no other wine that carries great age with such power and dignity. [CAMPBELL: *This is Maurice at his best.He takes you along with him when he describes these two old bottled sherries.*

[*Dr. Williamson was a great character who affected a small waxed moustache which extended for several inches east and west into sharp tapering points.He was a gourmet, kind, informed and hospitable, and his little house, clinging to the side of one of Guildford's steepest hills, was a museum of miscellanies, and not a very tidy one, and his garden too was inhabited by strange and exotic plants as well as a tank alive with goldfish.He revelled in his international, social and antiquarian knowledge and enjoyed the acquaintance of apparently everybody from Queen Mary down to myself: being among other things a collecting numismatist, he accepted from me one of the polygonal threepenny bits when they were first issued and claimed to have influenced the symbolic representation of thrift on their reverse side.*]

But old-bottled Sherries are not for everyone, and we must say a word or two about the wines more commonly procurable. Each of the great shippers offers a gamut of different kinds of wine, some suitable for drinking before dinner or as an occasional glass, some suitable as a dinner wine to accompany the soup, some worthy to be offered as the dessert wine, as an alternative to Port or Madeira. This, perhaps, is the moment to draw attention to one of Sherry's peculiarities, which makes it an economical wine; whereas most other wines deteriorate very rapidly once they have been opened, Sherry will maintain its quality, and, in the opinion of some people, even improve during several days in the decanter. Professor Saintsbury enters a *caveat* against applying this doctrine to the finer Amontillados; he says he has known a bottle of 'Tio Pepe' to wither appreciably between lunch and dinner. I confess that I have not had the same experience; the kindness of many friends, including its shippers, have given me a very fair acquaintance with 'Tio Pepe,' and I have found it beyond reproach after several days in the decanter. I have never tried an experiment which he says was a common practice when he was young, whereby the remains in the Sherry and Port decanters were combined in a composition called Matrimony. I doubt that the flavours would marry satisfactorily.

The wine that is usually served as a 'short drink' is Amontillado. This label, like other Sherry labels (and indeed one cannot limit this criticism to any one wine), has been monstrously abused. The bottle of Amontillado that you will get if you send out to the nearest public-house will contain a dreadful concoction, the shipper of which will have preferred to remain anonymous. But if you get an Amontillado that has been shipped by one of the well-known shippers you should get a light, dry wine, with a very clean, and sometimes a walnutty, flavour. 'Fino' means fine; but there is a dark mystery that I myself have never quite solved about the use of the word in connection with Sherry. Usually it means an Amontillado of more than ordinary delicacy; the La Riva 'Fino Macharnudo' is an excellent example of this, and Ivison's Amontillado Fino is another excellent example of a delicate light wine, with a delightful fresh flavour. 'Tio Pepe,' which is by some people thought (erroneously) to be a type of wine, is an Amontillado of great fineness, but it is nevertheless a stouter wine than those just mentioned; as most people are aware, it is shipped by Gonzales, Byass and Company. I think it is Williams and Humbert that ship 'Pando,' another delightful Sherry of great finesse, and Pedro Domecq has a 'La Ina' of a similar high quality, although I confess I do not find its flavour quite so much to my liking as the others I have named. I should have thought that 'Pando' was a Manzanilla; I may be wrong, and it is certainly very agreeable. Duff Gordon's 'Coquerino' and Ruiz's 'Isabelita' may also be specially commended.

These, however, are rather expensive wines, varying from 8s. to 10s. a bottle, and we cannot all afford to fill our decanters with wines that cost so much. And price is not decisive; you may have a lucky palate, that likes a cheaper wine, the cost depending on the value of the wines that have gone to the composition of the Solera, and a large proportion of fine old wine necessarily sending up the price. But they used to have (and may still have) at the 'Senior' an Amontillado that used to be called 'the 7d.,' because that was the modest price of a glass; I believe it was supplied by Christopher's, but I never learned who was the shipper. I only know that I invariably asked for it in preference to the more expensive Sherries which my host would press upon me; and I believe I was sharing the opinion of most of the knowing members of the club. And for many years I was never without an Amontillado which I believe was shipped by Duff, Gordon and was bottled by J. H. & J. Brooke under the name 'Natura.' About five years ago something put it under a cloud for a moment; I was told that during the troubles just before the Spanish war some discontented workmen misbehaved themselves and put dirty straw in some casks, from one of which I got a parcel. I am happy to say that the eclipse was only momentary, and it is now our standby on the Midland Circuit. For what might be called a 6s. Sherry (which I think is about the price at which one ought to get a good wine for regular use) it is difficult to better. There is an excellent Williams and Humbert wine at sixpence dearer that can also be relied upon to give satisfaction.

These latter not-finissimo wines make an agreeable accompaniment to the soup, and can be carried in from the ante-room. But some people like a fuller wine with the soup, and Barry Neame is very fond of serving a José Ramirez of a Solera 1858, which certainly matches his giblet soup to perfection. But this is a very rich wine, to go with a rich dish, and probably a Vino de Pasto, a wine less dry than an

Amontillado, but by no means luscious, will best fill the bill. But there is no label which in this country so often announces a detestable content in the bottle. 'Vino de Pasto' only means 'Wine of Repast'; but the grocer is satisfied that it means 'Wine mixed with Paste,' and apparently boot polish at that; a wine to be offered to the man who likes something with a bit of body in it, eh, what? And the chemical factories at Cette and elsewhere supply him with oceans of this detestable liquor, wherein the Sherry, if indeed any genuine wine has gone to the making of the brew, has been obliterated by a foul, earthy-tasting substance that would have been much more usefully employed as manure. I can never understand how such things continue to be sold; but it is called Sherry, and therefore Aunt Matilda and the Reverend Cluppins think it *must* be good, especially as Mr. Gumblebottom the grocer is such a nice man. But you don't get things like this from a good wine merchant. [CAMPBELL: *Most wine merchants sell a Vino de Pasto which does not merit Maurice's castigation. It is generally a good, sound, drinkable wine.*]

Amoroso and Oloroso are two types of sweet Sherry, the latter generally sweeter and fuller than the former. These wines are made full and sweet by the addition to their recipe of a larger quantity of *arrope*, which, as we have seen, is a Sherry that has been concentrated by the process of boiling it until it has been reduced to two-thirds of its original bulk (I believe the *arrope* is still further reduced to two-fifths to produce '*color*' for the purpose of darkening and strengthening the wine); and the Pedro Ximenes grape produces a wine far too sweet to be drunk itself, and only to be used as a flavouring, which is sometimes used as it is, but sometimes reduced to a concentrated form, called '*dolce*,' which goes to sweeten and strengthen the Amorosos and Olorosos. Some people like a glass of one of these with a rich soup, but they go better in my opinion with dessert; walnuts have no better accompaniment.

I am not quite clear about either 'Old Brown' or 'Old Gold.' I have usually found that 'Old Brown Sherry' has been a rather sticky and very inferior Oloroso; my experiences of 'Old Gold' and 'Pale Golden' have been pleasanter; I presume that the colour implies an absence of any '*color*' or perhaps even *arrope*, the strengthening being achieved by '*dolce*' alone. Incidentally, I was once privileged to taste the pure wine of the Pedro Ximenes grape; it was so intensely sweet as to be bitter. Don't laugh at that paradox until you have put a tablet of saccharine in your mouth; you will find that this sweetest of substances will taste almost like alum.

'East India Sherry' is a title that can mean little to-day. It used to be the practice to send Sherry by sailing-ships to the Indies and back; the masters were glad of the casks as ballast, and the rocking of the boat seems to have matured and improved the wine. Professor Saintsbury tells us that Mr. Walter once tried the experiment of mounting two casks of Sherry on the printing presses of *The Times* and keeping them there for six months, when he bottled them as 'East India Sherry' and 'West India Sherry' respectively. His guests would gravely affect to detect a difference of flavour. But I imagine that the modern presses of Printing House Square run too smoothly for such an experiment; and the vibration and hot pipes of a modern hold could hardly be expected to have a beneficial effect on any wine, so that we shall hardly hear of real or artificial East India Sherry again.

Sherry ought to be drunk very cold; tepid Sherry is horrible. The refrigerator does it no harm; do not put ice *in* it, or in any other wine.

And Sherry has one other quality, which, alas! to-day must be its most widespread appeal. It can tolerate tobacco, and no other wine can. So the chain smoker need not abandon all wine; he can still try to redeem himself by sipping his glass of Sherry. [CAMPBELL: *Maurice has handled sherry remarkably well, although obviously not always quite sure of his ground: in actual practice the traditional solera system is simple, sure and successful.*]

STILL WHITE WINES (I)

WHEN the soup has been removed and the fish makes its appearance Sherry has played its part, and a larger glass requires filling with a wine that can remain unperturbed by piscine flavours. Most fish are strong and pronounced in taste, and to savour a really delicate wine in their company would be impossible. Other more discreet denizens of the water may be accompanied by wines which normally would be out of place at this course; it perhaps will surprise the reader to know that it is possible to drink Claret with salmon, one of the most distinctively flavoured of fish; while sole and turbot are always well-mannered and inoffensive, and cause no wine, red or white, to leave the table. But these may be served with a sauce that perpetrates the offence that they themselves have avoided; the anchovy has a knack of turning up as an accompaniment to the most innocent sole or turbot, and so completely monopolizing the palate as to make all discrimination in the matter of wine impossible. For such occasions Heaven has allowed large quantities of white wine to be made in Bordeaux and sold under the name of 'Graves,' although much of it comes from other parts of the Gironde. It certainly justifies Heywood's jest that 'none but sextons drink Graves wine.' To make it fall bright recourse is had so freely to the device of sulphuring the casks (a necessary and proper practice when performed in moderation) that each glass tastes as though it had been stirred with a lucifer match. There are well-known vineyards that owe their fame entirely to the white wines they produce; I have patiently searched out samples of their wine of a good year, duly chilled the bottle, and tried the wine as favourably as I could. I am bound to say that the distance between the best Graves and the worst Graves that I have ever drunk would require a micrometer to measure it. [CAMPBELL: *Too severe: there are other drinkable white Graves besides the Haut Brion.*] I make one honourable exception. It is not generally known that the great Château Haut Brion, where one of the most illustrious of Clarets is produced, also makes a small quantity of white wine. This is a true Graves, for it comes from the Graves district. And it is a very fine wine indeed: the only white wine of Bordeaux that I have ever tasted which I should venture to offer in competition with a white Burgundy. In a good year it has a bouquet that suggests Sauternes; yet it is absolutely dry. In a Soho restaurant I once heard a lady ask for 'a dry Sauternes,' which then sounded to me like a remarkable contradiction in terms. For I thought of Sauternes in terms of Château Yquem, most luscious and lovely of liquors; I had not then met the Haut Brion Blanc. Now, although the French do serve the rich Sauternes with the fish, I have always thought it a heresy; the flavours clash and destroy one another. But white Haut Brion can go very well with the fish; it pleases the nose with its rich aromatic bouquet, while the austerity of its dryness enables it to look with scorn upon the fishiest of dishes. I have even had the luck to drink vintages of this wine against one another; neither of them very great, for 1918 was not a good year for either red or white wines, and 1922 was only a little better for white than for red, for which it was very poor. But the two Haut Brions showed very prettily, and each wore its rue with a difference. Those sufficiently eccentric to desire my society will find a bottle of Haut Brion Blanc a good bait for the trap.

My rage against the average bottle of Graves arises from the thought that the soil which went to its growing might have been devoted to a red wine, wherein Bordeaux, like Sylvia, 'excels each mortal thing.' But happily there are other districts where the white wines come nearer to my palate. Of all these Burgundy is *facile princeps*. It cannot be too often proclaimed and reaffirmed that the white wines of Burgundy are as far (nay, much further) above the white wines of Bordeaux (not counting the Sauternes) as the red wines of Bordeaux are above the red wines of Burgundy. And there is such a range of them. Let us begin, in true Irish fashion, with a wine that is not, strictly speaking, a Burgundy at all:

Chablis. [CAMPBELL: *Chablis* is *a Burgundy, but many white burgundies claim to be Chablis without any right to do so. The Chablis area is very small, but its wine is very popular, one could almost say too popular.*] The district whence our Chablis comes (or *ought* to come) is away in the north, near Auxerre; practically all the wine we know as Burgundy comes from south of Dijon, and between that town and Chablis is a long stretch of country, undistinguished by any notable vineyards. Kimmeridge in Dorset has given its name to a belt of bituminous clay that appears to dive thence beneath the channel and to make a very limited reappearance at Auxerre. It is apparently this clay that gives the peculiar virtue to the soil of Chablis and causes the clean and delicate flavour of its wines. For Chablis is unlike the other white Burgundies, although made from the same grapes. The noble Pinot gives it its excellence, when it has it; not every bottle that cries 'Chablis! Chablis!' shall merit commendation. For, as elsewhere in Burgundy, there is a war of the vines. For centuries the Pinot has been in a sort of competition with the upstart Gamay, a vine that, like many of the base-born, can flourish where a more delicate plant would perish. Now, the Pinot will grow on the slopes around Chablis, but it declines on the lower levels, where, however, the prolific and thrusting Gamay will thrive. By the Law of Place-names of Origin, wine made in the Chablis district can be sold as Chablis, no matter of what grape it is made. But only the wine made from the Pinot is real Chablis, as a wine-lover understands the word. And, as French law does not run in England, the bottle of Chablis you get in a little Soho restaurant may not even be Gamay Chablis; it may come from one of the devil's-cauldrons where they make the wine that ignorant people drink to their harm.

Your genuine Chablis has a glint of green underlying its yellow, or so it always appears to my eye; it is lovely to look upon as to taste. It is probably entitled to the honour of having been my first introduction to great wine; for, as I have explained, I was a teetotaller when young. And my first lapse was produced by a piece of casuistry that I almost blush to record. In the summer of 1910 I was in Francis Meynell's company when we met my own parish priest, the late Canon

O'Mahony of Blackrock, known to a host of friends as 'Father O'James.' Most hospitable of men, he promptly swept us off to the Villa Villa, then at the height of its popularity and, as I discovered a few nights ago, now, alas! no more. Food was quickly provided; drink was discussed. And to my amazement the Canon, a gold badge of the League of the Cross on his watch-chain, suggested a bottle of Chianti. 'But,' said I, 'you and I are teetotallers.' He looked at me gravely, but with a twinkle in his eye. 'We took the pledge,' said he, 'for one of two purposes, that is to say, either we wished to avoid becoming drunkards or we wished to give example to others. Now, we can't give example, good or bad, here because nobody knows us; and the third of a bottle of Chianti couldn't intoxicate anybody!' I confess I prefer my own defence of my abandonment of teetotalism; but God bless the Canon, whose kindly temptation of that day was only one of a thousand things I owe to his affectionate care and guardianship. He sleeps amongst the people he loved, and the poor do not forget him. But he made it necessary for a logical person like myself to put my principles in order; and I decided to be in future what I called for want of a better title a territorial teetotaller. [CAMPBELL: *How he enjoyed this joke of his!*] That is, I kept the pledge rigidly in Ireland, but took what was put before me elsewhere. (And that pledge I kept until 1923, when my mother, a woman of imperious character, ordered me to drink wine at her table; and I could not disobey my mother.) But the result of Canon O'Mahony's sophistry was that when Francis Meynell and I were being entertained at The Hague in the autumn of 1910 by the Chief Justice of the Supreme Court of Canada (and I believe that Sir Charles Fitzpatrick is still alive, although he must be of Nestorian age by now) I was enabled to drink at lunch the delicious Chablis of the Club of which he was an Honorary Member. After twenty-nine years I can still remember the clean glint of that wine, and my surprise at finding it so unsweet, for I thought of all wines in terms of sweetness in those days. I wish I could have noted the particulars of the growth; it was assuredly something excellent, probably a Vaudésir, or perhaps a 'La Moutonne,' for those were the days before 'La Moutonne' had suffered eclipse. Whether it has yet been brought back to its pristine standard I am unable to say, for I have not drunk it for many years. But I had a correspondence with Monsieur Long Depaquit some years ago, in the course of which he assured me that he was doing all in his power to restore the ancient excellence of his wine. 'La Moutonne' was a little portion of his vineyard, not shown on the cadastral plan and therefore not subject to the Law of Place-names of Origin; in the old days the label meant that the wine was made from grapes grown on that particular favoured acre, but I could not get from the proprietor a clear statement whether the label was still so limited or whether he claimed the right to apply it to wine made from grapes grown elsewhere. For some time past the lists of Chablis published by writers on Burgundy have usually set Vaudésir at the head of the column, a place formerly occupied by 'La Moutonne.' My advice to those who drink such wines is, to judge them on their merits. Having mentioned these two vineyards, it is only fair to draw attention also to the excellence of 'Les Grenouilles,' a wine of great delicacy.

For every bottle of genuine Chablis produced there are sold probably a hundred bottles that are frauds. Browning is largely responsible. There are few students who have not chuckled over his monstrous rhyme:

> 'Then I went indoors, brought out a loaf,
> Half a cheese and a bottle of Chablis;
> Lay on the grass and forgot the oaf
> Over a jolly chapter of Rabelais.'

Thence comes their familiarity with the name; 'What shall we drink?' 'Oh! why not a bottle of Chabelais?' '*Not* Chabelais; Chablis.' 'Have it your own way.' And, incidentally, I protest against accompanying bread and cheese with a bottle of Chablis. It should have been beer, even if that spoiled the rhyme; but if you *must* have bread and cheese with wine, you must drink something rough and coarse, like a cheap Côte Rôtie or a very *ordinaire* Claret. Chablis, if good, deserves a more delicate background; and, if bad, is not worth drinking. And if he was lying on the grass, Browning was drinking his Chablis on a hot day, un-iced; a horrid thought. Chablis needs the attention of the ice-pail, as do nearly all white wines.

Without desiring to refine too much, I am not to be taken as suggesting that cheese and Chablis are hostile to one another; I merely say that Chablis is not the wine for a bread and cheese lunch. For a meal of that sort you want a good full cheese of the Cheddar kind, and the real cheese for Chablis is Gruyère. Even Brie and Camembert are too violent for it.

So much for Chablis. Let us now come down to the Burgundy proper district. Here in almost every commune famous for its red wine you will find a white wine worthy of your attention; but the most famous are the wines of Montrachet, Meursault and Pouilly. So great is the fame of Montrachet that the commune in which it is grown, formerly Puligny, is now Puligny-Montrachet. *The* Montrachet comes from a vineyard owned partly by Baron Thénard and partly by the Marquis de la Guiche; it is a rare and almost unprocurable wine, but, of a good year and properly matured, it is the greatest dry white wine in the world. [CAMPBELL: *I agree, but for safety's sake qualify by addition of the word 'beverage': 'the greatest dry, white, beverage wine in the world.'*] It is a glorious and majestic thing, and I personally would never care to see it served with any but the most discreet of fish courses; chicken is its better background. It has remarkable delicacy of flavour, but it has more body than Chablis; more body, without being fleshy. It is the body of an athlete; indeed, if ever the fortunate proprietors thought of bottling it under their own label I can think of no happier design for the latter than a reproduction of the bronze statue called 'Athlete with staff' in the National Museum at Rome. Chablis, on the same lines, might be represented by Antinous; and there is a seated statue of Ares, also in Rome, that might similarly stand for the best of Meursault. For Meursault is indeed a fleshy wine; there is something rich and almost oily about it that differentiates it from all other Burgundies. It has not lost its athleticism; but it has not got the clean-cut figure of a Montrachet. The latter suffers its name to be in part used by lesser vineyards adjacent, excellent in themselves without quite attaining to the highest rank.

These are the Chevalier, the Bâtard and the Demoiselles. All these add the name Montrachet to their titles, and thereby claim affinity with the chief of the clan; but a bottle labelled Puligny-Montrachet is only a wine from the Puligny district, with no more claim to merit than the soil of the commune gives it. Chassagne has also added the name of Montrachet to its communal title, as a part of the great vineyard comes within its limits; so that a bottle labelled Chassagne-Montrachet is probably only a district wine of the commune. By the way, Montrachet is pronounced Mon'rachet. And a great many people in this country may have been agreeably surprised by a wine that calls itself a bastard, when drinking the Bâtard Montrachet shipped by Louis Poirier, which has become very widely known in recent years. The 1928 was a lovely wine in all its phases; it is hard to find now, but the 1933 promises to keep to the same path, having already fulfilled that promise in its early stages.

The white wines of Puligny and Chassagne are lean and slender, but without any lack of strength; in Meursault we find a rich golden wine, quite as dry as other white Burgundies, but with a delightful, lazy oiliness about it that suggests Brutus rather than Cassius. [CAMPBELL: *I cannot remember finding a 'lazy oiliness' about Meursault though perhaps one could say it was normally fatter than Montrachet, though the latter is no Cassius!*] And whereas it is usual to warn the wine-lover against a bottle labelled with a fancy title, such as 'Crême d'Amour,' or 'Gloire de Dijon' (for the roses of Burgundy make excellent garlands but no wine), in Meursault we find a vineyard the wine of which is prettily named 'Goutte d'Or' that is in fact one of the best products of the commune (although only classed as a third growth), and the wine lives up to its fancy name. I can remember a Goutte d'Or 1915 drunk in 1933 that was amongst the first ten bottles of white Burgundy that I have ever drunk. Certain ladies can afford a little *embonpoint;* it only adds dignity to their charm. There is a gracious hostess that I know, who shall remain anonymous, for she fasts and abstains on more than the days commanded in order to get rid of that little extra layer of fat that in fact distinguishes her; and I silently couple her name with every good drop of Meursault that I drink.

A little further south one meets the twin names of Pouilly and Fuissé; and Pouilly may be taken as the generic appellation of this district. Here are wines of grace and charm that do honour to Burgundy; they are the runners-up of the best Chablis; and they are as clean and delightful a white wine as can be found, with a fumosity that especially commends them. Fumosity is a word that seeks ancient lineage; but I use it, here and elsewhere, as depicting a particular bouquet and flavour that suggests that the vintagers have ground flints into the must. The resultant wine is dry to a degree; it has a suggestion of gunpowder about it, no bad quality in these days; but through these more or less adventitious flavours comes the clean perfume of the white Pinot grape, and the host who orders a Pouilly and gets what he asked for will not blush for the wine. It was a Pouilly at the Windham that first opened my eyes to the excellence of white Burgundies.

There are a great many other good white wines of Burgundy, but some remarks that I shall have to make about the red wines apply also to the white; the label is never as reliable in Burgundy as it is in Bordeaux, and, whereas the amateur can play with Bordeaux with confidence, and back his own opinion without much fear of disappointment, in Burgundy he would be far better advised to trust implicitly to his wine-merchant, provided he has selected the latter with discretion. Walk through Soho, and in every little wine-shop you will see advertised for sale bottles of Montrachet, Meursault, Corton, Chablis and so on, at prices that ought to cause you to rub your eyes. And if you went in and bought one bottle of each and poured the contents into a series of glasses you would not be able to tell which glass came from which bottle. But try a similar experiment with the wines offered you by your own wine-merchant and you will be agreeably surprised by the individuality of each.

I cannot remember ever having tasted a white Beaujolais, and I am told there is none. But we come to some noble white wines when we get further south. The valley of the Rhône has a more reliable allotment of sunshine than is enjoyed by the Burgundy district, and vintages are more regular. There are three well-known names to be met here, Hermitage, Côte

Rôtie and Châteauneuf du Pape. All these names originally signified vineyards, and have come to mean a kind of wine. The original Hermitage is supposed to have taken its name from the retreat of a returning Crusader, who brought home with him from the Syrian wars some roots of the Syrah vine, which he planted near his cell. These prospered, and in due course he found himself drinking excellent red and white wine. The soil was ferruginous; and, lurking in every glass, a flavour of gunflint gave character and distinction to the draught. Professor Saintsbury hesitates between a red Hermitage of 1846 and a Romanée Conti 1858 when seeking to name the greatest wine he ever knew; but he says that the Hermitage was the manliest French wine he ever drank. And manliness is the note of Hermitage, both red and white. Like other wines, it has suffered from the curse of the phylloxera, and it is nowadays impossible to drink wines as venerable as our grandfathers knew, when sixty years was no uncommon age for a red wine. Through the kindness of Mr. P. Morton Shand and his father I once drank a Château Grillet that was over seventy years old, and delicious; but Château Grillet is not, properly speaking, a Hermitage, although a wine of very similar type. It was that fountain of charitable kindness, Charles Walter Berry, God bless him, that gave me my best Hermitage, both red and white; and there is no nobler wine, albeit other wines may have more charm. But Hermitage has an Irish link; there is a Clos Saint Patrice there, which is supposed to be the site of a resting-place of Saint Patrick on his way to Ireland. It is now the property of the former owner of the Festival Theatre at Cambridge; and it is producing a red wine that would almost convert Hitler to Christianity. Great, rich, glowing red wine, with a mouthful of bouquet to every sip, if you know what I mean; and yet with no side about it. It was Mr. C. Harold Ridge who thus thrilled me, and as I had lunched that day with Messieurs Brooke, Bodle, my palate was somewhat exhausted. Yet I well remember the wonderful freshness and exuberance of that Hermitage, and I discerned a reason for the sudden and overpowering success of Saint Patrick in Ireland.

Châteauneuf du Pape records the exile of the Popes to Avignon. It took Saint Catherine of Siena all her time to persuade them to return to Rome, and the irreverent insist that the wines of Avignon were the moorings so difficult to abandon. Côte Rôtie pays tribute to the constancy of the sun in the Rhône valley; the wines are on the harsh side, but they are full of sunshine. I am more familiar with the red wines than with the white, but I have drunk the latter locally with considerable pleasure. [CAMPBELL: *I have never tasted white Côte Rôtie and doubt if they are often shipped to this country.*]

While it is true to say that all these wines lean more towards the produce of Burgundy than to that of Bordeaux, it is a solecism to treat them as a mere subdivision of Burgundy. They lack the roundness and fruitiness of the wines of the Côte d'Or; but they have an individual quality, a sort of Navajo cast of beauty and majesty, that belongs as well to the white wines as to the red. The white wines are pebbly and clean; they are dry without being harsh or bitter, and they have a fumosity of their own. Rub two bits of flint beneath your nose and you will know what I mean.

If we went still further south we should find oceans of wine being produced, but nothing of distinction, although Seyssel tastes good in its native Savoy. But the Loire gives us wines of much greater character; and there is no lover of France who has not at one time or another gladdened his heart with a glass of that most grapy vintage, Vouvray. Vouvray always seems to me to be the opposite number to Mosel, and both must be judged by the standards set for a simple village maiden rather than by the more exacting rules of polite society. In both districts one finds a sparkling wine which, while unpretentious, is agreeable, and far more so than the foaming monstrosities that some Bordeaux proprietors produce as a blasphemy against the goodness of God. But Vouvray usually means a still, or perhaps a *pétillant* wine; drunk icy-cold on a hot day, with its delicious fragrance of fruit, it is both a treat and a tempter, for it is distinctly heady. I have drunk it with oysters, and have been surprised that so sweet a wine could be so agreeable a companion. But it is a very good form of 'elevenses' with a biscuit; it is a good luncheon wine, and a glass with the fish at dinner is a change, and is far from unpleasant. For some reason, although you can drink excellent Vouvray in Paris, it is difficult to get it good in London. Whether the salt air of a channel crossing affects it, or whether English cellars are unsympathetic to its genius I do not know; but the phenomenon remains. On the other hand, a little lower down the Loire we find an excellent wine made from the Muscadet grape, two kinds of which I have cellared to my complete satisfaction; it travels well. It is a harder and dryer wine with a strange flavour entirely its own, that is very hard to describe; it is almost as though the wine had been bottled in a room where there was a lot of wood-smoke. Then there is Pouilly, not to be confused with the Pouilly in Burgundy, which thinks itself blessed by God in the possession of some of the best wines in the world; and why talk to such people of wines that we consider better when those that they have are quite good enough, and very pleasant drinking at that? In particular, there is a Pouilly Fumé that has a remarkable flavour of its own, of which Mr. Berry has given me some very delightful samples; and this was probably the wine of Rabelais.

Then we cannot neglect the still wines of Champagne merely because we have grown to associate the name exclusively with sparkling magnificence. After all, it is hardly three hundred years since the invention of corks enabled Champagne to discover its true function in life; before that, there had been the keenest competition between the wines, red and white, of Champagne and Burgundy. And ten years ago when the world slump set in, and nobody could afford to look as though he could afford Champagne, the shippers of that wine bethought themselves of their pristine glory and once again invaded the market for still wines. None of these made any reputation. [CAMPBELL: *There was a still Blanc de Blanc champagne that made what might be called a bubble reputation in those years, but there was not enough to go round.*] And I was sorry to find, as I suppose was necessary in the circumstances, that there was no revival of red Champagne. I have drunk red Bouzy that seemed to me to stand up well to a light Burgundy; indeed, there is a delightful red Burgundy from the little Pommard vineyard of Les Epenottes which is bottled by Louis Poirier and is unlike any other Burgundy I know; and I found the red Bouzy very like this. Pellucid, fragrant, round-flavoured and entirely free from that touch of heat that suggests an addition of warmed wine, these bottles were full of pleasant draughts that brought no regrets on the morrow. I am writing a few hours after having partaken of a bottle of Burgundy, classic in its vineyard and its vintage, but not quite so aristocratic in its shipper; and I am afraid I have a slight malaise, hardly severe enough to be called a headache, but nevertheless a reproach to the wine. You may justly say that I shouldn't have drunk Burgundy in September; I confess the error of taste, but point out that this hardly defends the bottle.

However, the white Champagnes that I have drunk suffered no such reproach. They go with oysters better than their sparkling brothers, and almost as well as a Chablis. They support any fish dish; and they are a pleasant accompaniment to white meat of all sorts. They never pretend to be princes of the blood, and this is really their disadvantage, for they cost more than the average bottle of white wine, and when one exceeds the usual price one expects to get a University accent in the wine. Latterly, with returning prosperity, I have not seen them about so much, and I suspect they have taken up a more useful situation in a more exuberant form. But they did give me the chance of demonstrating the superiority of nature over art. I put a bottle in a Sparklet syphon and offered the product as Champagne. Nobody would believe me that it was in truth Champagne, although treated in a somewhat unorthodox way. So if you do want to drink sparkling Champagne, buy a well-known brand; home industry does not suit the wine. In justice to the Sparklet syphon, I have always found its effect upon a good drinking water to be exhilarating and soothing at once.

This leaves only one other district of France that produces great wine (except that perhaps I should have mentioned Sancerre, from Berry, that is like a pleasant variant of the Loire wines) and that is, the Jura. Hence came Pasteur; and here he found the wines upon which to carry out the elaborate study which led him to declare that wine was the healthiest drink for mankind. Arbois produces both red and white wine, the red being rather a russet than a ruby, and never pretending to the greatness achieved by its white sister. For it is here that we find Château Chalon, once a single vineyard, but now giving the charity of its name to the best wines of its neighbourhood. And these wines are in all probability the ancestors of Sherry, for, alone of the unfortified wines, they produce the *flor*, as Sherry does; they have the flavour of the finest of Amontillados; and they last without the assistance of Brandy to ages unattained by other unfortified wines. They are unusual wines in England; but Philip Trotter, that amateur professional, has made a speciality of them and sends out a representative list of them from the Mutilated House. I think it was a bottle of his that Mr. Douglas West gave us as a prelude to a delightful lunch at the Swedish Club, near the Port of London Building, last April. It was a 1929 wine, and beautifully fresh and cool; it had not the body of a Sherry, but it had all the charm of the floweriest of finos. It is not a cheap wine, but it is drunk in little glasses, so that it probably does not cost a great deal extra in the long run. Not a white wine for the fish course nor an after-dinner wine; there is a *vin de paille* from the Arbois that can worthily fulfil this latter function.

And so, as the primary object of this chapter was to help in the finding of a wine to follow the Sherry, let us break off here and turn to the consideration of other still white wines that have another function to fulfil. I am not overlooking the fact that there are white wines of Italy; Orvieto, now sweet, now dry, offers varied charms; Suave, a wine from Verona, is not unlike a Graves; Chianti, in its white form, can be anything, but never very much; Lacrimæ Christi is a Neapolitan blasphemy. And I have drunk white wines in Hungary and in Rumania that were not contemptible. But these rarely appear on an English wine list, and are certainly not good enough for special importation. I have never tasted Russian wine, nor is it likely now that I ever shall. So let us leave it at that.

STILL WHITE WINES (II)

THERE are people who expect a glass of white wine to be able to stand alone; to be treated with the reverence offered to a great Claret or red Burgundy; to be sipped and discussed with disception and qualification; to be praised in its passing and in its memory. There are only three white unfortified wines that can stand up to this test; they are Sauternes, Hock and Tokay. Some of those who know me best will smile to see Tokay included, for I am not a Tokay fan. But, just as Brahms means very little to me beyond an exuberance of sound that I classify as wallpaper rather than a picture, so do I recognize that in wine, as in music, I must defer to more educated opinion; and those who know have always set Tokay on a pedestal. To Hocks I am not always fair; for my own part I think I would rather have a bottle of Mosel than any but a remarkable bottle of Hock. But Sauternes has my heart. Sauternes at its best can set the most gracious of all crowns on a banquet; and, except at its very worst (and then probably because it is an impostor) is always agreeable and winsome. I am convinced that the miracle wine of the Marriage Feast at Cana was a prophecy of Château Yquem. For there is no liquid that has ever lingered in a glass which concentrates so much virtue, nay, perfection, as an Yquem of a great year. It must be a great year; once, having, as I thought, at last discovered a wine that would lure my beloved sister from her teetotalism, I took her to Kettner's and gave her a bottle of Château Yquem. It was a 1918; I had met the wine in a bottle of 1906 (glorious thing!) and the 1918 tasted like an ordinary Sauternes. And my sister remains teetotal. Had it only been the 1921 her downfall would have been assured. For I cannot believe that there has ever been anything quite like the 1921 Château Yquem. They used to say that the 1869 was the greatest Yquem ever made. I once drank it, at the Château, against the 1921, and I thought that the younger wine had all that the elder could offer and charms of its own besides. The only other time I had an opportunity of tasting the 1869 was at the Hind's Head at Bray, and the bottle failed us, being, of course, promptly replaced by Barry Neame with a delicious bottle of the 1921. [CAMPBELL: *I was present on the occasion. The* 1869 *was part of a remnant of sundry bottles and Barry Neame only brought it out 'on the offchance.' It failed us. Good though the* 1921 *Yquem is to-day, it is not, in my opinion, a patch on what the* 1869 *was at the same age and continued to be for another thirty years.*] I wasted no tears on the 1869.

Now, all these three wines share one peculiarity in the making, although not all Hocks are prepared so carefully. The grapes are allowed not merely to ripen, they are allowed to begin to shrivel with ripeness before they are pressed. When picked, they are set like the Sherry grapes to sun themselves further; then the best grapes are selected and pressed. In Tokay these are allowed to ooze their juice without pressing; this precious drip makes the Tokay Essence, which is the real Tokay. The wine made from the more ordinary grapes receives from one to five hodfuls of the used Essence grapes to each vat, the number of hodfuls being indicated on the bottle of the wine so made as 1 puttig, 2 puttig, and so on. The Essence is allowed to ferment separately, and when bottled has the peculiarity that it must be stored standing up. Apparently there is something in the wine that is hostile to corks. There is certainly something that is hostile to disease; Walter Berry's stories of the miraculous cures it has effected are a challenge to the sceptical. The ordinary wine has something of this merit, but to a lesser degree; it is rich in phosphorus, and is supposed to be valuable when December weds May. I have drunk the wine, and even the Essence, as an after-dinner wine, but never with great pleasure, and, in the case of the Essence, with a qualm of conscience, for we were wasting a precious thing. But I was hoping to gratify my mother. She was living in London at the time and was not enjoying the best of health. One day, as we were chatting, she said, a little wistfully, that she had never tasted Tokay. It was the first time in her life that I had ever heard her express a desire for a pleasure of her own. I dashed in to Christopher's, and I think that 'Annie' Irish sent around to Walter Berry to rob his precious store; at any rate, when we were finishing dinner a few nights later I was able to take around a bottle of the 1868 Essence. My mother declined it. 'Mother!' I cried, 'this is your Tokay!' So she grudgingly allowed me to fill her glass. A few minutes later I saw her take it up and sip it doubtfully. She looked at me. 'Is this your Tokay?' she asked. 'Yes, Mother,' I replied. 'Well, I don't think much of it,' and she handed me her glass. So failed my only attempt to give her anything she really wanted.

Of Hocks my experience has been considerable, but I am entirely inexpert. My great friend, Harry Huggett, took up the study of Hock with the same enthusiasm that I brought to the study of Claret; and I enjoyed the benefit of most of his experiments. He wrote for the Sette of Odd Volumes a little essay entitled 'Rhenish,' which, until the publication of Major Rudd's admirable volume in Constable's Wine Library, was the best book I ever read about Hocks. And we drank the wine together in Hamburg, in Berlin, in Leipzig, in Frankfort, in Homburg and in London; and I trust he drinks his in Paradise to-day, but I find the wine has lost its savour. Yet I freely admit that Hock has more than Meredith's authority for its claim to senatorial rank—'there are senior Hocks.' The wine ages with a certain amount of dignity, although I always find in it a vulgarity that I seem to discern everywhere in the German character. [CAMPBELL: *In spite of the 'vulgarity' Maurice professes to find in German Rhine wines, as in the German character, he pays them a good deal of complimentary attention in the following pages which are some of the best in the book.*] It seeks to conquer you by violence. It can truly claim to be the grapiest of all wines; but the grape flavour has lost the simplicity of its vineyard days, and has acquired a *hauteur* that renders it more admirable than lovable. A Hock always affects that part of my palate that reaches up towards the back of the nose; and it causes reactions there that I get from no other wine. They are unfriendly, although distinguished; they seem to say, 'I have condescended to visit you, but I perceive that you are neither "von" nor "zu" and I reserve my intimacies for such as are.' And when a Hock grows old this acridity increases, so that it recalls a magnificent princess, eternally conscious of her rank, perpetually on the look-out for slips on the part of those around her, holding her position by her dignity, but never commanding the affection of her *entourage*. A Mosel might well be the tire-woman of such a Hock; the same pride is reflected there, but she is conscious of her own inferior rank, and unconscious that this makes her more charming. She will jest with you, and dance a polka or barn-dance, and will not be too proud to pocket a few marks for the favour. Sometimes (e.g. Berncasteler Doktor 1921) she will make a rich marriage and meet the princesses on their own level; but she will not be very happy there, and her thoughts will always be back amongst her fellow-villagers, dancing on the fair-green, or walking through the woods with the sweethearts of long ago.

Let us speak first of the Rhine wines. They are divided into three principal groups, of which the most important is the Rheingau; Rhenish Hessen, including the vineyards along the banks of the River Nahe, is the second, and the Palatinate is the third. The wines of the Rheingau are usually esteemed the most delicate; those of the Palatinate are the richest and most luscious. Heresy pursues me through the vineyards; just as my taste runs to the red Graves in Bordeaux, prizing them above the Médocs, so do I here give my preference to the Palatinates. When the young Mozart slipped on the polished parquet at Schönbrunn it was the good-natured Marie Antoinette that broke through the stiff etiquette of the Court and picked him up and soothed him. With a child's heedlessness of the world of fact he said to her: 'I like you, and when I grow up I shall marry you.' A nameless grave awaited the one; a second Calvary the other. I sometimes think of the incident when I drink a Palatinate. I cannot imagine a Marcobrunner or a Steinberg coming to my rescue like that; but a Forster Kirchenstück would do it naturally and gracefully.

The best stretch of the Rhine from the tourist's point of view is that which flows from Mainz to Coblenz, where the Mosel comes in from the left. The traveller so proceeding has left behind him the Hesse and Palatinate districts; but, as he goes northward, he leaves on his right the confluence of the Main on which stands the little town of Hochheim, from which it is generally believed we christened the Rhine wines Hocks. Three years ago Mr. Alfred Mayer made the interesting alternative suggestion that English visitors to Germany would probably have been impressed by the oft-repeated toast phrase, 'Hoch soll er leben (thrice repeated), Drei mal Hoch!' and might well have so associated it with the wine as to give it as a baptismal name. Ingenious as the theory is, I confess that I should have expected to find some evidence in support of it if it were well founded. In his paper which I have mentioned, Harry Huggett pointed out that the wine was always 'Rhenish' in English literature until the end of the seventeenth century, when we find 'hockamore' taking its place. It need hardly be pointed out that 'hockamore' is Hochheimer, the wine from Hochheim, which, as we have observed, is on the Main, not on the Rhine.

A few miles below Mainz our traveller will find himself passing Rauenthal, a little town just south of Wiesbaden, which gives its name to considerable quantities of local wine, some fair, some very good. The river, which has been flowing due north, here makes a left-hand bend and for twenty miles flows practically due west, until it reaches Bingen, where it turns north-west again. The northern bank of that twenty-mile stretch rises steeply to some hundreds of feet; facing due south, it has been terraced and planted with the finest vines; and in rapid succession the traveller will pass by Marcobrunner, Steinberg, Hallgarten, Schloss Vollrads and Johannisberg, five of the greatest names in the vocabulary of German wines. For this stretch of the river-bank is called the Rheingau; and it is almost universally reckoned the brightest jewel in the cluster of German vineyards. Charlemagne is supposed to have noticed that the snow melted earlier on these hills than elsewhere; more than one of the vineyards claims to have been founded by his care. Schloss Johannisberg, formerly the property of the Bishop of Fulda, was given by Napoleon to Marshal Kellermann in 1807, just after Jéna; the Marshal presented it to Prince Metternich in 1816, and it has remained in the possession of his family ever since unless, indeed, Herr Hitler has altered this. It must be remembered that there is the same difference between a Johannisberg and a Schloss Johannisberg as there is between a Margaux and a Château Margaux; the great estate has given its name to a district, and a bottle labelled only 'Johannisberg' is only a bottle of table-wine. Even a bottle of Schloss Johannisberg will vary in its contents, according to the colour of the seal, there being about six gradings of the grapes, and, as is usual in the case of German wines, the casks bearing individual numbers. The great rival of Schloss Johannisberg is Steinberg Kabinet, a wine labelled 'Steinberg' without addition being a table-wine. The Steinberg estate has for a long time been the property of the Prussian State, and one must give honour where honour is due and confess that wine production is one of the matters that this strange people does very well. A little further up the river the Marcobrunnen estate proudly sets its own standard of excellence; I have not often come across this wine, but in 1928, as I passed down the Rhine under the shadow of its vineyard, I drank an excellent bottle of Steinberg Kabinet 1921, procured on the river-boat for 9 marks!

Let us examine the labelling question before leaving the Rheingau. As we have seen, the great Johannisberg and Steinberg estates have given their names to the surrounding vineyards as well, and a wine with no other description may be taken to have come from some humble neighbour of the great estate. Steinberg itself is in Hattenheim, and therefore its neighbours *ought* to describe themselves as Hattenheimers; but I fear that the custom is more honoured in the breach than in the observance. However, let us take a bottle labelled 'Johannisberger: Mittelhölle'; that means that the wine comes from the Mittelhölle estate in the Johannisberger district. But even then it may be the least good cask of wine made on that estate; we therefore look for the words 'Auslese' or 'Spätlese,' the former meaning that the wine is made from selected grapes, the latter from grapes that have been late-gathered. (I have never seen a bottle labelled both 'Auslese' and 'Spätlese,' though I cannot see why not.) A 'Kabinet' wine is very specially selected. Sometimes one finds the proprietor's name on the bottle with the word 'Wachstum' or 'Gewächs' or 'Crescenz.' This gives the owner's guarantee that the contents are as the label states them to be, and is something like the château bottling of Bordeaux.

The best vineyards of Johannisberg are, first, the Schloss, and next Klaus Johannisberg; then the Mittelhölle and Unterhölle, the Doef, Steinstuck and Steinacker. In Hattenheim, after Steinberg, the Nussbrunn is the best-known vineyard; then comes Ehrbach, the home of the Marcobrunner wine. Rüdesheim is the next district, and Rüdesheim Berg is a common label; but there are three Bergs, Rottland, Halpfad and Burgweg, so that an unspecified 'Berg' should put one on inquiry. The Schlossberg is probably the best-known vineyard here. Winkel boasts Schloss Vollrads, and the Neuberg vineyard is well known also. In Rauenthal the Nonnenberg and Kesselring are probably the most familiar labels; the Eltville wines are not common here, but Hallgarten, with its Schonhell and its Geisberg, has made many conquests. The Doosberg is probably the best-known wine from Oestrich; Kiedrich gives us a Grafenberger and Mittelheim an Oberberg. I never remember seeing any of the wines from Geisenheim or Eibingen or Lorch; while Hochheim is best known for its Domdechanei and its Holle.

The chief geographical point in Rhenish Hesse is Worms; and a diet of Worms, in its liquid form at least, need not be unpleasant. In the first place, we come across here the most familiar and most misused label of all the Rhineland, for it is here that we find Liebfraumilch. Who has not at one time or another grimaced over a bottle so labelled? And yet, when the label tells the truth, it covers one of the loveliest of the Rhine wines. For, strictly speaking, it should be a wine made from the grapes grown in the vineyards adjoining the Church of Our Lady (Liebfrauen Kirche) in a suburb of Worms. Wine labelled in addition 'Stiftswein' or 'Kirchenstück' will probably be genuine; and Messieurs Deinhard with an honest frankness such as one would expect from their distinguished firm label their Liebfraumilch as 'Hans Christof Wein' in addition, making clear that the vineyard from which it comes, although a neighbour, is not one of those already mentioned. Shun like poison a bottle labelled 'Liebfraumilch' and nothing else. It is a cousin of those you find labelled 'Pommard,'

'Beaune,' 'Chablis,' 'Saint Julien,' and so forth, and probably has come from the same source.

In the Rheingau nobody would 'desecrate the soil' with a rival to the Riesling grape; but in Rhenish Hesse the Sylvaner has at least equal popularity, and, in the case of the inferior wines, it greatly predominates. The Sylvaner stands to the Riesling here almost as the Gamay stands to the Pinot in Burgundy; and it must be remembered that the Pinot and the Riesling are believed to be the same grape. And just as the despised Gamay (although, it is true, a cousin, and not the same grape as the Gamay of Burgundy proper) burgeons into glory in the Beaujolais, so does the Sylvaner come into its own both in Rhenish Hesse and in the Palatinate. In the latter district the wines of quality come from the Riesling and the Traminer; but the wines of quantity come from the Sylvaner.

The two best-known districts of Rhenish Hesse are Nierstein and Oppenheim; and Nierstein is almost as Judas-like a word as Liebfraumilch. But Nierstein Domthal will probably please; and Harry Huggett used to praise the Auflange wine. Of the Oppenheimers, the Goldberg is the most popular label, but the Herrenberg has also a good reputation. The wines of Bingen never greeted me with any welcome, although Harry used to class as such a Scharlachberg, which I think was strictly a Budesheimer, of which he thought highly. The Nahe flows into the Rhine at Bingen, and some good vineyards have blessed its course; but Nahe wines are usually grouped with those of Rhenish Hesse.

The third great division of the Rhine wines takes its name from the Palatinate, a district situate south of, or above, Worms, about fifteen miles in on the west side of the river, enjoying the benefit of a tableland in the Harz Mountains. In the Oberland and the Unterland quantity rather than quality is sought, and here it is that the Sylvaner grape predominates; but in the Mittelhaardt are to be found some of the loveliest of all the Rhine wines, and in my view they show a clean pair of heels to the best of the products of the Rheingau. In the rich vintage of 1921 they excelled, for their nature is to be sweet and rich, and that glorious sunny year filled all the grapes with sugar and traced with the hand of a loving artist every line that went to the making of a perfect Palatinate wine. I think I liked the Forsters best; a Forster Langenbohl 1921 at the Windham Club was, if not the best Hock I ever drank, at least the Rhine wine to which I look back with most affection. Many a good friend of mine it sent on his way rejoicing in the good old days before the bottom fell out of the world. It was a magnificent wine, but it was friendly. It seemed to beg you to tell it about Château Yquem; was that wonderful creature so *very* much more beautiful? And you found yourself all but charmed into the lie of a negative. It had a great partiality for cold lobster; now, who could imagine a wine from the Rheingau tolerating the insult of a fishy-tasting dish? But perhaps it was at its best when you sipped it after dinner, with a pear to rob it of its overwhelming sweetness. It did not object to a bottle of Claret afterwards; but Port it refused to meet. It was lovely and pleasant in its life; and in its passing it brought a blessing and left it as a last legacy.

Harry was fond of the Langenbohl; he drank many a bottle with me, and returned them with others of the same rank at the 'Rag' and in his own house. But he liked the Kallstadters best of the Palatinates, and named in his paper the Annaberg and the Kobnert as the Rhenish wines that in the vintage of 1921 had pleased him best of all. But before leaving Forst, honour must be paid to the great wine of the Jesuitengarten, so much flattered by forgery and imitation, and the Kirchenstück, another majestic wine. And then we just glance at the districts of Durkheim (its Feuerberg is a favourite) and Deidesheim (where, perhaps, Kieselberg is the most likely choice) and Ruppertsberg, with its Goldschmidt and Schlossberg, and say farewell to the most winning of the Rhine wines.

Perhaps this would be the best place to make mention of a wine that has affinities with the Rhine wines, but comes from a different district and is strongly individual. That is the famous Steinwein, the vineyards producing which are situated around Wurzburg on the Main. It has always held to its peculiar, flat, flagon-shaped bottle, which makes it very difficult to handle commercially, and for this reason it is uncommon in England. But it is a wine with great merits, some of them secret and only to be mentioned to the wise. In a good year, such as 1921, it can attain great finesse, and it is always a wine of brisk body. It has the colour of a Mosel rather than a Rheinwein, pale amber, with a greenish glint; it is capable of attaining great age. It has a slight smack of Tokay about it, but whether this is a phosphorus content or not I do not know. It usually coincides with Hocks and Mosels in its good vintages; and a dozen, chosen in a favoured year, would give you the opportunity of offering your guests something unusual and pleasant.

When I think of the Mosel wines I am always reminded of that delightful operetta, *Lilac-Time*, in which the author had the happy idea of telling a story about Franz Schubert and setting it to that composer's own music. It was a brilliant success, and must have given pleasure to millions. And yet it would not seek to be classed as anything more than a light-hearted trifle, albeit good music and good art. And it well exemplifies a wine which only very rarely presumes to greatness, but which, when genuine, is always agreeable. In England we nearly always spell the name 'Moselle'; why, I do not know, for it is only when the river has crossed the frontier into German soil that it begins to nourish notable vines. In recent years much publicity has been given to the wines of Alsace, and some of them can be quite agreeable ordinaires; but they have a long way to go before they can compete with the wines of the Mosel. [CAMPBELL: *They have undoubtedly made progress in public favour; if they continue at the same rate they will become serious rivals of Moselles.*]

Mosel is a light, greenish-yellow wine of low alcoholic content, a lighter edition, as one might say, of a Rhine wine, but entirely free from the vulgarity which I have attributed to the latter. Mosel should be drunk young; four or five years old is about the best age. And it should be drunk very cold; a tepid Mosel is a horrid thing. It goes admirably with oysters; not quite as well as do the white Burgundies or Champagne, but much better than any other wine. It is never out of place with a fish course; but it can go on through a lunch or dinner, and, although not very suitable with either butcher's meat or game, it is quite unobjectionable in such company, while it is excellent with a roast goose, the greasy texture of which seems to be dissolved by its clean attack. At a meal where it is proposed to serve only one wine, Mosel is not a bad choice. Its tall, green, steeple-shaped bottle lends a grace to the table; by the way, I have never discovered the origin of the convention whereby Rhine wines are imprisoned in brown glass and Mosels in green. But the fact should not be forgotten, as it once was by a distinguished connoisseur, who destroyed himself in my presence by proclaiming the Mosel we were drinking 'the most delicate Hock he had ever tasted.' I had observed the green bottle!

The Mosel wines are made from grapes grown in the valleys of not only the Mosel itself but also the Saar and the Ruhr; and I hope that the vineyards are not at the present moment paying too dearly for Herr Hitler. The vineyards are invariably set upon the hillside, which is terraced to retain the soil around the roots of the vines. This terracing fulfils another purpose as well; because of the northerly situation it is important to catch all the sun that can be trapped. The stone terraces reflect the heat which they themselves receive, and the best properties are those which have the most southerly aspect. Here again we find that the principal vine planted is the Riesling. I am told that when German chemists were paying more attention to making wine and less to making gas and high explosives there grew up in the Mosel area a practice of cutting little elder-flowers and infusing them in a wine spirit. With the tincture thus made a flavour resembling that of the muscatel grape was given to wines which otherwise would have proved uninteresting. Like most dishonest practices, this recoiled upon the heads of those who sought to benefit by it; the taste for highly-flavoured wines vanished with the money of the profiteers who liked them; and for a time the pure Mosels lay under the bad name earned by these spurious wines. Since the return of a genuine and discriminating taste in wine honest Mosels have come into their own again.

Fermentation is Nature's most cleansing process, and I do not attach much importance to the protests of German wine-shippers that the human foot never touches the grapes of the Mosel. The wine would be none the worse for a contact which is traditional, which lasted elsewhere until days within living memory and which made some of the best wines the world has ever known. If a dirty sheet goes into a wash-tub that is properly managed it comes out clean. No impurity would survive the mad boiling of the must during fermentation; at the conclusion of the chemical process all that has been, so to speak, washed out of the liquid settles at the bottom of the vessel. But let me not be understood as despising this averment of cleanliness; on the contrary, all good habits are to be praised, and this one gives some guarantee that, after the wine has been made, it will be preserved with care and protected from contamination. For this reason one welcomes the boast of the Mosel wine-growers that their vineyards are the cleanest in the world, although one may smile to learn that every visitor must wipe his boots before passing between the vines. I wonder how they treat their manure before spreading it!

The Mosel vintage may not begin until the local committee have given their approval; and only during the hours they prescribe is work in the vineyards permitted. Any over-eager proprietor who continues his labour after the church-bell has rung the signal for him to stop will be fined a sum varying according to the magnitude of his trespass, the fine going to the communal poor-box. The grapes are unloaded, when gathered, into a sort of gargantuan coffee-mill; here they are crushed, but not pressed, and it is a mixture of berries and juice that subsequently goes into the hydraulic press, there to be squeezed until the last drop of juice has been extracted from the murk. This juice, or must, goes immediately into casks, wherein it undergoes its fermentation. In an unfavourable year an absence of sunshine may have left the fruit deficient in sugar; the owner may at this stage have his must analysed, and may then add pure cane sugar to make up the deficiency. If he adds too much he offends against the excise laws and runs the risk of having his wine confiscated. The inferiority of Art to Nature is always demontrated by the wines so assisted, which are never the equals of those made in the good years without the addition of sugar.

Mosels, like the Rhine wines, are fermented in casks, or *fuders*, each of which has a number; and individual casks will vary to an extraordinary degree. For this reason one frequently finds the more exclusive wine-merchants selling these wines under a label which carries the cask number as an additional indication of origin. The fermentation takes place at a temperature artificially maintained at between 80 and 90 degrees Fahrenheit; and, as the bungs of the casks are left loose to allow the gas to escape, the sound in a fermenting cellar would remind one of a turkey farm, the perpetual 'gobble-gobble' sounding almost like propaganda. When the fermentation is complete, the ullages are made up, the bungs are driven home, and the casks are removed to a cool cellar, where during the next thirty months the wine will be racked some six or seven times, by which time it will be ready for bottling. If drunk young, as is advisable, it will be found that there is still just the least sign of a continuing fermentation, which causes a slight prickle on the tongue, and makes the wine seem spritely.

The labelling of Mosels follows the same practice as that which obtains on the Rhine. The name of the District comes first, and the name of the Vineyard in that District comes next. To take a simple example, a wine from the Berncastel District, illustrious or humble, would be a Berncasteler; if the label said no more it would probably mean that the wine had come from some unimportant little vineyard in that area, or even that three or four little proprietors had clubbed together to make one wine between them, all their grapes being grown in the Berncastel District. If the wine is labelled 'Berncasteler Rosengarten' it means that not only have we a wine of the Berncastel District, we are assured that it comes from the Rosengarten vineyard in the Berncastel delimited area. If the year also appears on the label you will know that your wine is from a particular vineyard in a particular year; and the words 'Auslese,' 'Spätlese,' 'Wachstum' and 'Crescenz' mean the same here as on the Rhine.

Probably the Mosel that enjoys the greatest fame is the Berncasteler Doktor, a vineyard that belongs partly to Messieurs Deinhard and partly to the Thanisch family. The wines of this estate, like those of the Schloss Johannisberg on the Rhine, are bottled under seals, the colour of which varies according to the standard of the wine. The cheapest is yellow; then follow the lilac, pink, orange and white, until at last we reach the Auslese under a gold seal. The nearest competitor of the Berncasteler Doktor is the Piesporter Goldtropfchen, Piesport being the District and Goldtropfchen being the Vineyard. The Graach wines also enjoy well-merited popularity, the Himmelreich being perhaps the favourite. Zeltingen and Wiltringen are the only other Districts that need be mentioned in such a hasty survey as this; the Scharzberger wine from Wiltringen is worthy of serious attention.

Do not fall into the error, opposite to that under which so many people have laboured, that Mosel is a great wine, or to be compared with Rhine wines, to say nothing of the great French wines. Its proper 'opposite number' is Vouvray, a wine that always leaves pleasant memories, while never pretending to be a prince of the blood. And to quench a summer thirst, or to introduce those vintages too great for sudden drinking, it would be hard to find a more agreeable medium than a good bottle of Moselwein.

It would be ungracious to close this chapter without acknowledging my indebtedness to Mr. A. J. Koenen's little book *From Vineyard to Table*, in which he gives a most animated account of the making of the German river wines. I had read this before I read Major Rudd's book, already mentioned, and it gave me my first instruction in the mystery of Mosel. The bulk of it is devoted to the making of this wine, and the Rhine wines only receive a short treatment, on the ground that 'they are made as are the Mosels.' I confess I was pleased by this slight to an over-proud district!

CHAMPAGNE

IN the days of the Consul Plancus the City of Cork was a haven for all sorts of unexpected argosies, some of the most agreeable being theatrical and musical. Indeed, the late Sir Frank Benson saw to it that the young idea in Cork should be enabled to shoot along the most approved lines; his annual visits gave us three weeks of delightful education, and he was pleased to say to me half a dozen years ago that he had found the Cork audiences amongst the most critically appreciative he had ever faced. The Compton Comedy Company was another regular visitor; while to the musical stage came Arthur Rouseby, Carl Rosa and Charles Manners with their respective companies; and we were enabled to hear some of the most distinguished vocalists of the day. Nor were these our only loves. There was a sincere but unpretentious body called the Elster-Grime Opera Company that used to take the lesser stage of the Palace Theatre and give rattling good performances of the lighter Verdi and similar operas; after forty years I have a vivid recollection of a presentation of *The Daughter of the Regiment* that nothing at the Opéra Comique in Paris could banish from my memory. And suddenly we learned of a still more gallant effort; the G. Hilton St. Just Grand Opera Company had engaged the Assembly Rooms for three weeks, with G. Hilton St. Just himself singing all the tenor parts. The city buzzed with the excitement of pleasurable anticipation. We heard with delight that our excellent Lord Mayor, Mr. Augustine Roche, himself a keen patron of music and the arts, intended to give them a civic welcome; and, as Mr. Roche in private life was a wine merchant, we had no doubt that the company would be well entertained. Nor were our ideas erroneous. The reception took place at 6 p.m., and the performance of *La Traviata* began at 8 p.m. I had secured a good seat, and while waiting for the overture to begin I could hear that the Lord Mayor's hospitality had been transferred to the stage; the popping of corks was intermittent but frequent. But punctually at eight the orchestra appeared; it consisted of a pianist and a harmonium player; I think that sometimes a flautist joined them, when not engaged in singing on the stage. But courage conquers all; a creditable performance of the overture took place, and the curtain rose upon the revel of the first act. The caste were dressed in eighteenth-century costumes; but Mr. Roche's waiters lent a modern note, and continued to uncork Champagne with an assiduity and a hospitality that filled me with direful presages. Never was the drinking song rendered with greater gusto; the chorus splashed Champagne in every direction, and the principals sang with the chorus. Mr. G. Hilton St. Just, no mean tenor, was in excellent voice; and I felt that for once, if I was not getting exactly what Verdi had intended, I was undoubtedly enjoying the authentic atmosphere which Dumas had meant to create. But Nemesis was at hand. The volatile wine carried the company through the first act with tremendous *élan;* but when the curtain rose for the second it soon became evident that there was something rotten in the state of Denmark. When Alfredo's father came to reason with his son, the latter sat down at a table and buried his face in his hands; at the end of each line of the Di Provenza he lifted it, glared at his father, and buried it again; occasionally a sob of emotion (which looked uncommonly like a hiccup) shook his body. Meantime his father, who enjoyed a powerful bass-baritone, kept alternating between hissed pianissimi and thundered fortissimi, something like this:

> *Far a*WAY *in old Provence we have* WAI*ted thy return;*
> *We have waited* THY RETURN *far away in old Provence.*

By this time there was no restraining the gallery. ' 'Gor, boys! 'Tis great stuff Gussie's been givin' 'em!' cried one voice, and fifty others joined in a chorus of not unfriendly chaff, which did not affect the performers in the slightest. The opera finished triumphantly, Mr. St. Just's final 'O Violetta!' being a magnificent degeneration from a musical note into a hysterical sobbing. Never had Champagne a more agreeable triumph. And if I know my Gussie Roche, it was no mere non-vintage wine that was being offered; it was probably the 1900 vintage and a first-class brand.

I have told that story because I think it illustrates several of the qualities of Champagne. In the first place there is no wine which will work more quickly as a pick-me-up or a banisher of constraint. It is amusing to observe the frozen silence which usually attends the opening of a dinner-party, and which vanishes in a buzz of conversation as soon as the Champagne has been round. The late Horatio Bottomley had many demerits; but he was clever enough to appreciate the value of a good draught of Champagne at eleven o'clock in the morning. Harry Preston used to say that eleven o'clock was the ideal hour for the wine; just a chicken sandwich and a pint. For many years it was my habit to begin my celebration of Christmas by attending midnight Mass, and then partaking of a very early morning breakfast consisting of some of the delicious spiced beef of which Messieurs Russell of Cork appear to have the monopoly, home-made brown bread and butter and a bottle of Champagne. My subsequent sleep was peaceful, and I awoke, filled with Christian charity towards all men—or almost all. The very appearance of a bottle of Champagne on the table gives an uplift to the spirits; and anything that does that in evil days deserves a decoration, well represented by the gold-foil around the neck of the bottle.

Wherein lies the special virtue of Champagne? Surely in the fact that it, of all wines, alone knows how to deal with its own gas. The still wines burke the problem by discharging the gas in the fermenting vat and in the cask; the sparkling wines of the Rhine and the Mosel, still more dreadfully those of Burgundy and Bordeaux, seem to be boasting of their sacrilege. I am told that His late Majesty, King George V, liked sparkling Burgundy; it is the only thing I have ever heard of him that was not to his credit. And, after all, *de gustibus non est disputandum;* which of us does not nourish a secret passion for some kind of food or drink of which we are heartily ashamed? I myself—no; I won't tell you about it, and it has nothing to do with Champagne or sparkling wine. Sparkle has an attraction of its own; witness all the fizzy waters we find, the most popular being those in which the sparkle is most pronounced, as in Perrier water, or closest grained, as in Apollinaris, and even those who most despise a glass of fair water will not turn up their noses at a glass of soda-water. But we recently had ample opportunity to test the matter with Champagne itself. When the lean years came the bottom fell out of the market for sparkling Champagne, and many of the shippers began to send us a still wine which was cheaper and less showy. But it seemed to most people to be a mere white *ordinaire* from Burgundy, without distinction or attraction; and I rarely see a still Champagne on the list now. The sparkling wine has retained its charm. If our means permitted, it would be evident that it had retained its popularity also. I trust that in the present war its vineyards may be spared; in the last, three vintages were shipped, 1914, 1915 and 1917, the grapes for all of which had to be gathered under shell-fire and the wines matured, so far as Rheims was concerned, under a perpetual bombardment. I once made the suggestion that hoteliers should continue to mention these wines in their lists, even when the bins were empty, and even to print them with a little laurel wreath around the dates; for never were wines so gallantly made. And the blessing of heaven was on the 1914's and the 1915's at any rate; I don't think I ever met the 1917 myself. [CAMPBELL: *Lucky fellow!*]

But we were talking about the sparkle of Champagne. The sparkle is natural to every wine [CAMPBELL: *This may be true, but it conjures up horrible possibilities. All things may be lawful, but all things are not expedient! Let us turn a deaf ear to the tempter*], for it is merely the carbonic gas which lies combined with alcohol in the sugar of the grape; but the gas naturally seeks to escape, and, if left to itself, it will blow the bung out of the cask, and the cork out of the bottle. I cannot recall any evidence that the ancients were familiar with sparkling wine; familiar, that is to say, with sparkling wine as a beverage, for they must have been aware of the presence of the gas in the young wine, and I imagine they must have allowed this to work itself out before they sealed the wine into amphoræ. If they had filled the jars with young wine and sealed them with pitch or wax as was their custom, they would have found the wine sparkling when they removed the seal. And, as sparkling wine looks pretty, I cannot suppose the many poets who sing of the Roman banquets failing to notice the phenomenon. As I can recall no mention of such a fact, I assume that it was not present. After the Roman days, the practice of sealing the bottles seems to have been discontinued for many centuries; the usual stopper was a drop of oil, with a wisp of tow above it. In the sixteenth century some anonymous genius recognized the virtues of cork-bark as a stopper for a bottle; but I suspect that the first corks were conical in shape, and just shoved home. It is to Dom Perignon that popular legend ascribes the discovery of sparkling Champagne; and it may well be that in the course of his experiments he observed that the corks were being driven out of his bottles by the effervescing contents. He is then supposed to have tried the result of tying down the corks, in which case, of course, the gas would develop to its full capacity and the result would be something like our modern sparkling Champagne. But, as I have pointed out, the same result would have resulted with the wine of any district. Where Dom Perignon's real merit lay was in the blending of the grapes from the different regions of Champagne, of which the Abbey received large quantities from many areas in rent, giving in the result the one juice in the world that is really able to deal with its own gas and make a wine that is both brisk and pleasant, free from mawkishness and full of vitality.

The Champagne area has been strictly delimited; outside that area no vineyard may call its wine 'Champagne.' The privileged district runs south of Rheims past Epernay, down as far as Cramant and Avize. It is said that the grapes of Ay alone are themselves capable of making a wine that, unblended, can hold its own as a Champagne. Whether this be true I cannot say; but I do not think that any of the great shippers give their blessing to the legend. All of them prefer to pin their faith to a blending of the grapes from four different regions, each house having its own formula and recipe to determine the proper proportions. It comes as a surprise to most people to learn that Champagne is made from black grapes; at least, there are three black grapes to one white in the blends. They are the same grapes as make the wines of Burgundy; the difference is to be sought in the soil, which in Champagne is of a much chalkier consistency than in Burgundy. As you drive through the vineyards you will come upon a section of the road where the slope has been eased by a cutting. The sides of this bit of road will show you a kind of architect's elevation of the soil and subsoil. You will be amazed to find less than a foot of gravelly loam over a subsoil of more or less solid chalk. The poorer the soil the better the grapes—from the wine-grower's point of view; whether this would apply to dessert grapes I do not know; ask Mr. Middleton. And the closer the grapes are to the ground the better they are. They are certainly pleasant eating; and it is pleasant to observe that no stint is set to the munching of the gatherers. The vineyards are in part owned by the great shipping houses, and in part by private owners from whom the shippers buy. The process of gathering is much as in other districts; the pickers empty their small baskets into larger hampers, over which preside snippers who cut the larger bunches into smaller ones, eliminating any defective grapes as they do so. Nor are the hampers allowed to hold their burden for long; in every gathering area there is set up a powerful hydraulic press, to which the hampers are conveyed as rapidly as possible, lest the black grapes should commence to ferment before being pressed. If this were to happen, the tint of pink from the black grape-skins would be irremediably transferred to the wine; whereas, although the weight of the press squeezes out a pinkish juice, the subsequent fermentation gets rid of this, and the wine remains pure golden in colour.

I have only once seen a press at work; it seemed to consist of a sort of tray about 10 feet square and 3 feet deep, into which were thrown what I was told was 4,200 kilogrammes of grapes of one kind. Down came the piston; and the juice ran out to fill ten casks of 200 litres each. This first pressure produces what is called the Cuvée; the murk is separated and pressed again, giving what they call the 'first cut.' This second pressing, or 'first cut,' may go to the making of a first-class wine; the 'second cut' never does. Out of the murk remaining after the second cut they distil the 'Marc de Champagne' which can be very agreeable; many connoisseurs prefer it to the 'Marc de Bourgogne.'

Now, just as we had to hurry our grapes to the press lest they should start fermenting in the basket, so we must now hurry our casks of must into Epernay or Rheims lest they should burst on the way, through premature fermentation. Their destination is the cellar of the shipper; on their arrival, ten litres are drawn from each cask, and the bung is put in loosely. Now for the next month the must will boil and bubble like a witch's cauldron; and the space that had been occupied by the withdrawn ten litres will prevent spilling. After that month the first fermentation will be over; the wine will then be racked into clean casks three or four times, leaving behind in the old cask all the deposit and any cloudy wine. But not quite all the cloudy wine is left behind; for it is in the cloudy wine the ferments are, and a couple of wine-glassfuls of cloudy wine must go into the new cask or else all fermentation would stop.

The wine being now practically star-bright, the time has come for each individual shipper to exercise his own skill and judgment in the blending of the vintage he is going to offer to the public. All the casks are 'single': they are the unblended produce of one district. But by now the shipper has under his hand casks from every district, ready to be blended as he wishes. He must first make up his mind whether he is going to ship a vintage wine. He will only do this if the year has been a good one; and he has also to consider the state of the market, as his wine of the previous vintage may not be going as fast as would call for replacement. One of the biggest houses made a vintage wine in a fairly good year a few years ago, and suddenly decided that the state of the market would not warrant them in shipping it. So they bottled it as 'Non-Vintage'; and many purchasers must have been startled at the excellence of the wine they thus bought at the cheaper price. For a non-vintage wine is a blend, not merely of the various kinds of the Champagne wines of one year, but also of the wines of different years, some old, some young; and, like most artificial things, it is not as good as Nature's own effort. This would appear to be in contradiction to the result of blending the various wines of one year; here Art has undoubtedly improved on Nature; but when it comes to mixing wines of different years the story is a very different one.

There may be something in the suggestion that such a marriage needs the blessing of added alcohol, as in the case of the fortified wines, which blend so beautifully. However, there the fact remains; a vintage Champagne can nearly always be relied upon to give long odds to a non-vintage wine from the same house.

But even in a vintage year the shipper will not bottle off all the wine he makes, otherwise in a poor year he would be without any means of improving a wine deficient in one or other of the qualities that go to make a good wine. So what he does is this, he first will blend his vintage wine in vast oak containers, adding so many casks from this area, so many from that; when the blending is complete and the constituent wines have had time to marry he will re-cask as much of the result as he intends to bottle; and he will store the rest in great sealed glass-lined tanks or casks to play their part in improving the deficiencies of a less fortunate year. In the same way he will make up his supply of non-vintage wine to meet the demand of the market, and will re-cask that in a separate part of the establishment, ready to be bottled as and when needed.

Now we come to the really interesting and peculiar part of the preparation of Champagne. Up to this stage the wine is still, and it contains a small amount of matter calculated to dim its brilliance. At this point of its career it is bottled, and to each bottle is added a small spoonful of *liqueur de tirage* to assist its further fermentation. *Liqueur de tirage* consists of pure cane sugar, dissolved in old wine; it is cloyingly sweet and sticky, and a very little goes a long way.

When the liqueur has been added the bottle is corked with a temporary cork, which is only hooked down, not wired. We now make our way to a cellar where we find on every side what are called 'pupitres'; they look something like the boards of a sandwich-man, slotted to hold the bottles horizontally. Their bases are close together, and they look like a tall and narrow A. The bottles are inserted head first; every day a man visits them, and, taking the bottoms in his hands, shakes them with a twisting motion. This results in the sediment slipping down the sides and channelling along the underside of the bottle. At the end of the process all the sediment that has been marring the perfect clarity of the wine is gathered on the cork.

At this stage used to take place, and still takes place in some establishments, an operation calling for the greatest dexterity, and vital to the success of the wine. A skilled operator takes a bottle, eases the cork, and, as the gas sends it flying out, tilts the bottle upright, so that nothing but the cork and its load of deposit is lost; scarcely a drop of wine is spilled. In most establishments, however, modern science has been called in to replace the skilled operator; the bottle is set with its neck in a freezing mixture; when the wine in the neck is frozen the bottle is set upright, the cork is removed, and with it comes a little piece of ice containing all the deposit. In either event, the wine is now crystal-clear; the few drops of lost wine are replaced from another bottle set in a machine for that purpose; and it is at this stage that the wine is liqueured with *liqueur d'expédition* to fit it for its particular market. The bottle is corked and wired and sent with its fellows to mature in the chalk cellars until the time comes for its shipment. Perhaps you have guessed why Champagne should appear amongst us, all gay with silver or gold foil. This dates from a time when the skill of the disgorgers was not so great; and accordingly some bottles lost a little more than others, and might appear ullaged. But a discreet curtain of gay tinsel concealed such inequalities from the eye, and in effect nobody was a penny the worse. Besides, ullaged Champagne is, in the opinion of many people, improved; see what George Saintsbury says in his *Notes on a Cellar Book*. I have only once consciously tried it myself, and I did not form any concluded opinion about it.

The wine will lie in its own cellars for months and even years; and as long as it sleeps in the sympathetic chalk it will remain unimpaired. Furthermore, if it find a chalky home in England it will live equally happily, while in the neighbouring bins its cousins from Burgundy and Bordeaux will chafe and fret, and go up to the table in so irritated and angry a state as to be unrecognizable. Yet they too are having their lives lengthened in the chalk. I can illustrate this by two examples. When I visited Epernay in 1933, the Comte de Vogüe crowned my tour of the Moët et Chandon cellars by opening for me a *gamme* of their wines which went back from 1926 to 1884. The bottle of this earliest year opened with a 'pop' that would have done credit to its youngest brother. It was a little 'woody' in flavour, but, three years older than myself, it put me to shame. Had it been cellared in clay it would have been dead thirty years ago. Again, some six or seven years ago my late good friend Jim Brocklebank gave a little party to taste old Champagne. It was at a time when we were wondering what we should drink after the 1921's, for the 1928's had not yet spread their magnificence before us, the 1923's were very steely and hard, and the 1926's were nothing to make a song about. [CAMPBELL: *Maurice obtained most of his knowledge of the making of Champagne—as of other wines—from the most reliable sources, and his verdict that* 1923's *were 'very steely and hard' and* 1926's *'nothing to make a song about' must receive due consideration, but personally I am not of the same opinion. I found some of the* 1926 *extremely good.*] But I think the youngest wine we drank that night was the 1915, and we relied mainly on 1904's and 1906's. These had come up from a chalk cellar in Hertfordshire and were entirely admirable. Young Champagne, which is everybody's drink nowadays (or rather what those drink who drink Champagne at all), is a delightful young débutante that charms with many graces, but is entirely unaware of the serious side of life. Old Champagne has lost the first sparkle of youth, but has gained in winey-ness; it is a wise counsellor, none the less attractive because of its wisdom, but not to be treated as a mere sport of an hour; it is a wife rather than a mistress. And so those lovely old wines delighted our hearts, while a beautiful Mouton 1899 sulked upon the sideboard, demanding at least a month in a London cellar before it would put on its company manners.

The cellars in Champagne are of great interest. They appear to date back to the Roman occupation of Gaul, and they are witnesses to the magnificence of the national mind that invented the arch, and left us such triumphs of gargantuan architecture as the Colosseum. Slave labour was responsible for the cellars of Rheims; they are probably the quarries from which the Romans drew their building material. An opening, some twenty feet square at the surface, was dug downwards pyramidically until the hall below, at a depth of some sixty to eighty feet, measured thirty yards square. The workmen must have been raised and lowered by cranes. From these squares galleries were driven in every direction, and ultimately stairways were cut to the surface. On the hottest day in summer you will be well advised to take a warm overcoat with you, if you intend to stay below for more than a few minutes. [CAMPBELL: *How wise! this thought for his reader's comfort is characteristic of Maurice's sympathetic Irish nature.*] And here in the galleries are stored millions and millions of bottles of Champagne, like the forty thieves, asking each visitor: 'Is it time, Captain?' During the short period of the German occupation of the city in 1914 the greatest nervousness was not unnaturally felt by the shippers; but the Germans were afraid of some trap in the catacombs and almost entirely abstained from investigation, so that little damage was done.

It will be observed that Champagne stands about half-way between Claret and Sherry, inasmuch as it is not a perfectly natural wine like a Claret, and yet, although in a sense a blend, it is not as much the creature of its shipper's fancy as is a Sherry. But it is obvious that Champagne depends upon its shipper quite as much as does a Sherry; and it is therefore prudent to gain acquaintance with the names of the score of firms that can be relied upon to keep up a certain standard in their wines. The first name I learned to respect was Moët et Chandon; the brand has fallen a little in popular favour since then, but largely because the house refused to bow down the knee before the new gods of Hurry, and continued to make their wine for houses that had cellars where Champagne could quietly mature for many years. To-day a Moët of 1921 should be compared with another brand of 1926. I believe that in 1933 the house decided at last to adapt its policy to modern conditions, and I have no doubt that this will be followed by a return of popular esteem, as I personally have always liked the wine. Pol Roger is probably the brand that has most frequently found its way to my own cellar; I always think it is underrated in England, and is fully the equal of higher-priced wines. Krug holds my allegiance as the king of them all; my recollection does not go behind the Krug

1919, but that was truly an excellent wine. And Krug 1928 must be the best wine made in the present century. The 1929, I am sorry to say, disappointed me; but probably only by reason of the contrast with the 1928. I never thought so highly of the popular 'Widow,' and Bollinger has not seemed to me to justify its premium; but P.J. (as Perrier Jouet is popularly called) is nearly always a brisk and cheerful wine, well worthy of distinguished attention. We had a lot of Ayala on the old Munster Circuit; it was the 1904 wine, and I am afraid that we did not rate it highly. For this sentimental reason I was delighted to find the Ayala 1928 so excellent the other day; quite able to take care of itself in any company, and to take care of the company, too. For personal reasons I have a *tendre* for Charles Heidsieck; the Dry Monopole and the Piper are unfamiliar to me. Lanson used to be a very pretty light wine; except the excellent 1928, I have not tasted it since its 1920; and I believe the house did not meet with approval in the trade when it put its 1933 on the market a few years ahead of everybody else, and the other houses have now decided not to ship that vintage. Irroy is another wine of which I have lost sight; I have enjoyed it. I have drunk delicious Pommery; but its 1915 was an experiment in dryness that caused a damp on its popularity, which even the excellence of its 1926 did not altogether remove. At a recent dinner I was sitting next a well-known rival shipper and they served the Pommery 1929. I had difficulty in persuading him that it was not the 1928. Higher praise for a 1929 there could not be. Recently I have observed an increased demand for the Duc de Montebello, a wine that would have to be really bad before I would refuse to love it for the sake of the gallant Marshal Lannes. Professor Saintsbury did something to establish the reputation of Saint Marceaux south of the Tweed. And of course there is always Mumm; but in the case of the taste of a prattler like myself it is hardly to be expected that I should subscribe to the doctrine 'Mumm's the word!' Yet I have recently found its 1920 excellent. [CAMPBELL: *I have no quarrel with his selections, and of course he, nicknamed 'The Prattler' in the Society called 'Ye Odde Volumes,' must have his little joke about subscribing to the doctrine 'Mumm's the word!'*]

I have only named a few of the principal firms; and in

speaking of these I am not to be taken as suggesting that there are not many other houses where you can count on getting an excellent wine. Certain houses are fashionable, and demand and supply fixes a higher price in their case than elsewhere. Most reputable wine-merchants search out and find an honest little shipper who enables them to offer quite a sound Champagne at about 9s. a bottle where they have to ask 14s. or even 16s. a bottle for the more fashionable brands. But in the long run it will be found that the big houses have earned their reputation. And it may beget further confidence in them if I mention the fact that nowhere in the world are there to be found rivals that work together in greater friendship and harmony. Visit one of them and you will hear him loud in the praise of every wine except his own; and if he entertains you at a restaurant it will be the wine of another shipper that he will offer you. In Bordeaux every great house is connected by marriage with every other great house; but Rheims and Epernay seem to be on much more friendly and even affectionate terms than their southern cousins, where perhaps the neighbourhood of Gascony provokes a wit that sometimes proves Beatricean.

Because of its sparkle, Champagne has no competitors. This was not always so; in the days when wines were still it was the secular rival of Burgundy, over which it had the advantage of an easy waterway to the capital. Louis XIV by his own conversion swung opinion in favour of Burgundy; fortunately, this was on the eve of the discovery of the sparkling wine, so that the disadvantage did not last long. And Champagne in its new form soon enjoyed royal patronage; both Louis XV and the unfortunate Louis XVI liked it; poor Marie Antoinette was a teetotaller. Napoleon, although Chambertin was his favourite wine, which the wretched man shook up all over Europe and drank with water, had a penchant for Champagne also, and was a frequent visitor to Epernay, where he liked to visit the cellars, although he probably omitted this on the occasion of his last visit, which was during the campaign of 1814. Soon followed the occupation of Paris by the Allies: their officers carried the taste for Champagne all over the world; but in those days it was a sweet Champagne. Meantime, Scotsmen had been putting their money into distilleries and their sons into medicine; thence dates the absurd medical opinion that whiskey and soda is better for you than wine. Whiskey and soda was also a sparkling drink; but it was dry. It was about this time that the lady who then directed the fortunes of Pommery found a particularly persuasive young man named Hubinet at her ribbon shop, whom she sent to England to learn English and sell her wine for her. He said: 'If you will make a dry Champagne instead of this syrupy dessert wine people can drink it with their meals and you will sell it.' Hence dry Champagne, hardly sixty years old.

With the wave of prosperity that marked the latter half of the nineteenth century and the beginning of the twentieth, Champagne entered upon a period in which it well might have cried, 'Save me from my friends.' Every vulgarian, every roué sought to express his hospitality in terms of a wine, the finer points of which meant nothing to him, while the expense and the gold trappings meant a lot. From the account of its process of manufacture it is obvious that Champagne can never be a cheap wine; but cheap imitations can be made. Carbonic acid gas can be pumped into the most ordinary Algerian wine; the bottles and gold foil can be the same, and the labels can be made to look very convincing; but the subsequent headache will mark the difference. The taste, of course, would be different also, for all who possessed a discriminating palate; but taste would be the last thing to associate with the people of whom I speak. And in economics there is a principle known as Gresham's Law; base coinage will always drive out the genuine coins. So did the base imitation Champagne drive out the genuine article; for the bad reputation earned by the spurious wine drove away would-be customers from the true product. Champagne remained a wine; but 'Fizz' was an intoxicant; no occasion of factitious or artificial gaiety could be complete without it. And I am afraid that even I, jealous of the honour paid by undiscriminating people to this gracious wine, while they ignored my beloved Claret, have been heard to speak of Champagne words which I greatly regret, withdraw and blot out with a wholehearted apology.

Champagne must be treated seriously and as a great wine,

which it is. It is also the best-tempered of all wines; it calls for no coddling and resents no ill-treatment. Send a case to the North Pole and then to the Sahara Desert; bring it back to London, finishing the journey in the shakiest of conveyances; give it half an hour in the refrigerator, and it will pour into your glass, not, of course, at its best, but clear and well-flavoured and quite pleasant to drink. [CAMPBELL : *I agree, hence my disagreement with any treatment likely to make it 'not, of course, at its best.' Some people might take him seriously.*] 'No other wine can do that. A tired man can find no better pick-me-up; a thirsty man no more welcome quencher. The digestion that is too delicate to stomach other wines will find that the sparkle in Champagne is a digestive. It has a tonic effect upon the brain more rapid than that of any other juice of the grape; and though Champagne may go flat, no Champagne dinner ever does. It can be drunk right through the meal, although at a formal dinner I would prefer to serve it first and last, allowing fine red wines to fill the interval. It pleases every sense; and nothing is more charming than to see it served in those hollow-stemmed glasses, which cause it to send up a perpetual little fountain of bubbles in the centre.

But all this has to do with its sparkle; and Champagne must also be considered as a wine. And here it is capable of giving, although to a less degree, the same interest that can be had from a consideration of the finer Clarets and Burgundies. You must look underneath the sparkle; there is a wine there, made from the Pinot grape, with an infinite variety of flavour, some of it due to the nature of the wine, some of it to the skill with which it has been blended. Do not think, because some silly young people drink it out of a lady's slipper, that this is the proper measure of respect due to it. In health refrain from misusing the good creature, and in sickness you will be able to fall back upon the life-giving assistance of a trusty friend.

Some of the shippers continue to make for their own consumption still red wines of Verzenay and Bouzy, and pétillant white wines of Cramant and Ay. By the courtesy of the Marquis de Polignac I was enabled to taste these; and I suspect they owed some of their charms to the gracious hand that poured them. But they have vanished from the markets, and must be drunk in their own homes, as a gift. Nor are any of the gifts of hospitality lacking in Champagne. Shall I forget how for two days Monsieur Marcel Charles Heidsieck wrestled with my ignorance, and, not content with teaching me about wine by day, instructed me on the English Cathedrals by night; how we journeyed through the long hot days, my ears resounding with the praises of every other shipper's wines, until, the long journey ended, it was almost a surprise to find that the cool draught with which I was invited to restore my strength was a delicious bottle of his own 1926? Drink Champagne wherever you are, and think you know the wine; you will never truly appreciate it until you learn that all that is delightful in it is only a reflection of the lovable people by whose care it is made.

CLARET (I)

CLARET is wine. It is *the* wine. It is entirely natural. It is almost infinite in its variety. It is lovely in its colour, lovely in its fragrance, lovely in its flavour, lovely in its companionship. It is free alike from headiness and from heaviness; it is digestible itself and a good digestive as well. Champagne meets the tired man with boisterous good humour and swiftly animates his weary being; Claret is more reserved but not less sincere in its welcome; the sense of well-being comes more slowly and subtly, and lasts longer. Burgundy again differs in its approach; it is bluff and hearty, more solid than Champagne, but less intellectual than Claret. [CAMPBELL: *I welcome the word 'intellectual,' a tribute to himself and all those who put Claret first among wines.*] For that is the true secret of the greatness of Claret; it is essentially an intellectual wine. You drink your Champagne heedlessly, hardly troubling to identify the shipper; you take your glass of Burgundy, content to classify it as good or better. But set the most ordinary Claret on the table, and at once the conversation turns to growths and vintages, districts and bottlings; the little *ordinaire* is recognized as the cousin of Château Margaux, and appreciates the inquiries about its more distinguished relatives. You can always drink Claret with pleasure; but you can never drink it with real satisfaction unless you are prepared to talk about it. It is the wine of the *connoisseur* as well as being the wine of the *gourmet.* Of all the pages that have been written about wine, nine-tenths have been devoted to Claret. I think I am right in saying that it has been the favourite wine of most literary men, although some of them allowed their appreciation to take strange forms. Keats loved to cover his tongue with red pepper to intensify the beautiful coolness of his Claret. (But then Thackeray drank Burgundy with Bouillabaisse—'The Chambertin with yellow seal'!) Browning did not disapprove of Bishop Blougram's icing his Claret; and I fear that many others have overwarmed it. The Frenchman thinks no shame to drink it well diluted with water—a crime we rarely commit. But we, on the other hand, not infrequently dose it with various fruits and spices and drench it with soda-water, calling it Claret-Cup; while there is another vandalism called Mulled Claret which is happily dying out, in which the noble wine (for noble it is in its humblest form) is boiled in a saucepan with an orange or a lemon stuck with cloves, and drunk when the steam has been lit. Neither the orange nor the lemon has a flavour compatible with that of Claret, and a potation that wastes three good things is to be condemned. But, on the principle of 'Trial and Error,' it is to be hoped that some, who first encountered the taste of Claret in one of these heretical forms, improved upon the knowledge thereby acquired and subsequently became a disciple of the gentle wine. For Claret is gentle as well as noble, and forbears to make parade of its own splendour.

Claret is the red wine of Bordeaux. Occasionally one finds in old literature a reference to some other red wine under the name of Claret; but these are so rare that it may be stated in general terms that Claret has been the Englishman's name for red Bordeaux wine for hundreds of years. It obtained its name from the clearness of its colour. There are other wines which would probably have been shipped to England from Bordeaux, although coming from further districts, the colour of which was thicker and darker than that of the native wine. Their qualities were in all ways inferior to the wine of Bordeaux, to which the word 'Claret' was applied as a distinguishing compliment. Confusion as to the origin of the name has arisen from the fact that there used also to be imported a wine called *vin clairé*, which was a mixture of red and white wines; but one has only to look at the price lists to see that this was something inferior, and unrelated to the beautiful thing of which we are speaking. There was also a frightful confection called 'clarré,' which was a compound of red wine and honey and ginger and canel, and must have been about as clear as mud. Similarity of names has sometimes caused confusion as to the origin of the name of the most popular of all wines.

When we limit the application of the word Claret to the red wines of Bordeaux, we have to set a further geographical confinement, too detailed and elaborate to be set out in words, but immediately ascertainable by reference to the map of the Bordeaux area in *Bordeaux et ses Vins* by Cocks et Feret, a book that contains everything that is to be known about the Claret country. It is sufficient for our purpose to say that the main subdivisions include the Médoc, Graves, Saint Emilion, Entre-Deux-Mers and Palus. Of these the first three are the most important; I think I am right in saying that no great wine comes from either of the last two. The Médoc is usually given pride of place, although a much more modern wine district than Graves; while Saint Emilion, where the country is more broken and wooded, gives a more constant wine, its best years and its worst years being not so far apart as is the case in other districts.

Bordeaux has been fortunate above other wine areas in one respect. In Burgundy, to take a contrasting example, the great vineyards were in many cases the property of religious communities; and, when the anti-clerical laws came into force, these large estates were broken up amongst small proprietors, upon whose personal standard of honour the customer had to rely for the purity of his wine. Some were impervious to temptation; others introduced baser but more prolific vines into their little parcel of the great vineyard, and sold the resulting mediocre wine by the aid of the famous name that they were legally entitled to use. Bordeaux has been faced by no such problem. Its great vineyards have in many cases remained the private property of a family that continues to give hereditary care to their cultivation and improvement; while, in most cases where the property has had to be sold, a syndicate has been formed to buy it as a whole; often the original owners become shareholders; the vineyard continues as a single entity; and in most cases the ancient proud tradition is handed on as faithfully by the new owners as it had been by the old. For this reason the label 'Château Latour' has a ring of authenticity that the label 'Chambertin' could never possess. The Burgundy may have been made by any one of a couple of score of proprietors; it may be the finest wine of the vintage, or it may be a pitiful product of the Gamay grape, not to be drunk amongst Christians. But a bottle labelled 'Château Latour' contains a wine made from grapes grown on the Latour estate, all picked and pressed in the same manner, and no portion of the vineyard being planted in a different way from the rest. For this reason the amateur may venture to compete with the professional in the vineyards of Bordeaux, and buy according to his own instruction. The amateur who ventures to set his own opinion against that of a Burgundy shipper is courting disaster.

My dear friend, Barry Neame, who has shown that the best cellars need not be found in large places, has a Claret List that boasts over one hundred different bins of Château-bottled wines. At his hospitable table one may sit down to compare half a dozen vintages of the same wine, or the same vintage in half a dozen wines. It is his pleasant habit to make each guest 'mark his card'; and it is both instructive and amusing to find the extent to which opinions will diverge about the varying merits of the individual bottles. What is the secret of all this difference? That secret is to be found in the variability of several factors that go to the making of a first-class wine. These factors are (1) the soil, (2) the vines, (3) the intelligence and industry of the *vigneron*, (4) the amount of sun in that year, (5) the amount of rain in that year, (6) the times when the sun shone and the rain fell, (7) the presence or absence of such misfortunes as mildew, oidium and other plagues that afflict the vine, (8) the skill and intelligence of the *maître de chai*, who controls the making of the wine, (9) the care with which the wine is matured before bottling, (10) the judgment and skill with which bottling was done, (11) the subsequent history of the particular bottle: how often it was moved, was it kept in a good or bad cellar, and so on, (12) the care with which it has been brought up, *chambrée*, and decanted. [CAMPBELL: *By* chambrée *I think he really means 'at the temperature of the room'; he was no advocate of taking the chill off by warming the bottle at the fire, or of the equally pernicious practice of swilling decanter or glasses with hot water. The twelve factors enumerated by Maurice show a surprising knowledge of viticulture which one may justly envy.*] This is a formidable list; yet every factor here indicated goes to make or mar the success of the wine.

Soil, for example, is at the base of the matter. It is the difference of soil between the two places that primarily differentiates the wine of, say, Bergerac, from that of Bordeaux, not so far away, although there may be here subtle differences of aspect and temperature as well. It is the less noticeable difference of soil that makes Médocs such silky, gracious wines, and red Graves such noble warriors—the distinction always seeming to me to be like that between glossy and matt prints of photographs. Between the parishes of Saint Julien and Pauillac there is hardly a visible boundary, although if you look closely you will find a thread of a stream running between the vineyards of Léoville and Latour; yet there is *some* difference of soil, and Léoville is always recognizable as a Saint Julien and Latour as a Pauillac.

The vines of Bordeaux are four in number; but *the* vine is the Cabernet Sauvignon, which represents from 80 to 95 per cent. of the crop, the remainder comprising the Merlot, Malbec and Petit Verdot. At Château Mouton Rothschild, for example, the proprietor aims at making a wine that will be almost pure Cabernet, only about 5 per cent. of Petit Verdot being added to soften the extreme severity of the more austere grape. This used also to be true of that portion of the Château Lafite estate called the Carruades plâteau, where a second-rank wine used to be made in these same proportions and bottled as 'Les Carruades.' This wine grew progressively better until in that miserable year 1922 it outclassed the Château wine; since then it has always been included with the first wine, and is no longer made separately. But there are very few estates where these proportions obtain; more usually 15 per cent. of Merlot or Malbec goes to 85 per cent. Cabernet. The Merlot and Malbec grapes are softer and more perfumed, and add a suavity that many Clarets are glad to possess. The wines so made reach their maturity more rapidly than those made on the sterner formula; Mouton nearly always requires three or four years more than they do to reach a comparable stage. But the various vines do not always develop in the same way, and the *vignerons* can only plant hopefully, guessing at the proportions. In 1927 the Merlot vines flourished vigorously while the Cabernets hung back. The result was an ill-balanced wine that in some estates positively stank. One honourable owner of one of the finest vineyards in the commune of Margaux refused to sell his wine even as 'Margaux.' On the other hand, there were other estates, the owners of which were more fortunate in their crops, and I have seen some Clarets of 1927 in reputable lists, although I have not tasted any. Stop: let me be accurate. In 1930 there used to be a restaurant called 'L'Atelier' in the West End of Berlin, owned by a Herr Kahla, where one fed very well, and the wise drank a *pichet* of red or white wine, as the case might be. The red wine was announced as 'Bordeauxwein 1927.' I tried it and found it very drinkable, but much too robust for a Claret; so I challenged the smiling proprietor, who promptly surrendered. 'Mr. Healy,' he said, 'I will tell you, because you have guessed. It *is* a Bordeaux of 1927; but it has been blended with a Spanish wine of the same year and if we had not done that it would not have been palatable.' But I am still waiting to find a really good 1927 that I can brag about as the successor to my Château Haut Brion 1910, a wine of a mildew year in which no other vineyard, so far as I know, was able to show a wine, in spite of which the Haut Brion lacked little of the essence of a great Claret.

By the way, Château Angludet, an excellent vineyard in Cantenac, still favours the Carmenère, a variety of the Cabernet that is more luscious than the Cabernet Sauvignon. I do not know in what proportion the vineyard is planted, or whether the other vines are mingled as well; but some of the Carmenère vines at Angludet are over a hundred years old. [CAMPBELL: *It would have added interest if Maurice had told us if and how the wines of Angludet escaped the phylloxera invasion.*] I have met the wine in several vintages and have always found it individual and excellent; there is a je-ne-sais-quoi of perfume about it that is distinctive and pleasant, and it ages beautifully.

The next five of the factors that I have set out above deal with the growth of the grape. Industry has no fairer field, nor one in which the reward is richer. Baron Brane found the Mouton vineyard a neglected estate when he took it over;

by dint of loving care and incessant labour he turned it into an estate producing the wine that of all Clarets commands the highest price; but of course much of the credit must be given to the present proprietor who has continued upon the excellent lines laid down by the Baron. And de Brane was not contented with one such triumph; he bought a second vineyard in Cantenac, and, having brought it to the perfection of cultivation, gave it his own name. Château Brane Cantenac is still reckoned amongst the most successful of the second growths. On the other hand, I have myself seen the fall from grace of more than one great vineyard, formerly reckoned amongst those certain to please, but now trusting too much to Nature unassisted by industry. It would be invidious to name such; but at least one is a Saint Emilion. For vines need careful dressing, pruning, replacing and a dozen other cares; those who neglect the trouble will soon see the result.

Nature, too, can be very capricious. The grape needs protection from frost in its youth; a warming sun, but not a scorching one, both in the spring and in the late summer, to fill it with sugar; gentle rain to swell its bulk and give it plenty of juice; a freedom from hail. To get all these gifts in the right order means the creation of a vintage year. For frost may nip the vines and leave a miserable fraction to bear; rain may be so plentiful as to induce mildew; the sun may burn too fiercely and scorch the skins, in which case the wine will have a sort of burnt taste, something like the flavour of *crême brulée* (you will get it in most red wines of 1906, 1921 and 1926); the rain again may play the worst trick of all and drench the ripe grapes in September, washing away the bloom, in which lurk the little ferments in their millions, and without whose help no good wine can be made. A good wine can only be made when the vines have escaped frost and disease, when there has been a sufficiency of gentle rain at the proper times, when there has been a fine ripening sun in September, and no drench of rain at the end to destroy the ferments.

The actual making of the wine does not give quite so many opportunities for the exercise of skill; but there are moments. In the hot summer of 1895 the temperature of the air was so high that the must lacked oxygen and refused to ferment. The Abbé Gayon was consulted; he was visiting Château Lafite, and saw the sulky vats himself. 'You must make it all over again,' he said. And so the must was pumped out of one vat into another; the extra exposure to the air did the trick, and the result, in such vineyards as had been able to benefit by the Abbé's advice, was a wine of great delicacy, with the most remarkable bouquet I have ever found in a Claret. Elsewhere, the vintage was a complete failure. Here is a good example of the exercise of skill. But more often the grapes are crushed and fermented for a fortnight or three weeks and the wine is drawn off into its tonneaux without any departure from dull routine. But then commences a two-and-a-half-year care of the wine in the *chai* which calls for judgment and attention. Perfect cleanliness and a low, even temperature must be maintained unless you want to run the risk of your wine turning to vinegar; it must be drawn off from cask to cask several times, and must be fined with isinglass or some similar substance as well. Finally, when the time comes for bottling, a great deal of judgment is necessary. The course of the wind, the phase of the moon, the temperature are all matters of importance. (I do not know whether the phase of the moon is a superstition or not; but it is just the sort of thing that the so-called scientist would sneer at, without realizing that it held an undisclosed secret of great importance.)

Once Claret has been bottled, you begin to feel that it is past all dangers. Not so. In the first place, the difference between a good bottle and a bad bottle of the same vintage may be that one has lain still in the cellar since a few weeks after being bottled, while the other has been carried from pillar to post, and has never had a fair chance to rest. We know that a bottle of Claret throws a fine deposit, which movement will shake up into the wine; but this is cured by a few days' rest, and does not explain the undoubted fact that a wine that has been constantly moved, even when it has enjoyed a few weeks' rest before being decanted, is never as good as one that has remained in the one cellar all the time. And there are further complications. If you do not know what has happened to your wine since it left its Château you cannot guarantee that it has not spent half its time in a refrigerator and the other up against a radiator. Now, wine is a living thing; it is full of little organisms that are doing their work very quietly and unostentatiously, and when they die the wine dies too. When they are alive they like to be comfortable, and they are very susceptible to changes of temperature. Alternate boilings and freezings are no more to their taste than they would be to yours. So they may take it into their heads to start a hunger-strike, and die; and that is why Sir Moneybags Bullion's priceless Château Lafite 1899 has all gone to water, while yours and mine, having enjoyed a good home, are still in their prime.

But even now the obstacle race is not over. The bottle that has lain in the perfect cellar without disturbance has still to be brought to table. This calls for the greatest care. In the first place, see that the decanter is perfectly clean and bright and has not got a teaspoonful of the water of its last washing ageing at the bottom. Your bottle should be brought from the cellar at least twenty-four hours before it is going to be drunk so as to take the temperature of the dining-room (for my part, I prefer it to take just a shade more warmth—a cool corner of the kitchen often does it very well). There are two views as to how decanting should be done, to say nothing of those eccentrics who say that it should not be done at all. The first school say that as your bottle has lain on its side in the cellar, the dust will be channelled along the underside of the bottle, and that *therefore* (which nearly always foretells, as in this case, a *non sequitur*) it would be fatal to stand the bottle upright before decanting. They counsel you to keep the bottle horizontal in a decanting-basket, draw the cork laterally and decant in the cellar. I prefer to stand the bottle up for one or more days before decanting. In the first place you can look at any number of bottles and you will find that, for some mysterious reason, the dust does *not* channel along the underside; it seems invariably to confine its presence to the neighbourhood of the punt. Whether this comes from its having at some stage stood upright for a long time I do not know; if this were so I should have expected the corks to dry and shrink and the wine to be ullaged, and I have not found this to be so. Accordingly, I am firmly of the school that prefers to stand the bottle on its base for twenty-four hours or more and then to draw the cork vertically. Before drawing the cork, wipe the outside of it and the neck of the bottle with a piece of clean muslin dipped in cooking brandy. This removes any risk of the outer dirt giving a foul taste to the wine. Moisten another piece of muslin in the brandy, and when you have drawn your cork, wipe out the inside of the bottle-neck, and let the bottle stand for a few moments. [CAMPBELL: *I must confess I have never tried wiping the inside (or the outside for the matter of that) of the bottle with muslin dipped in cooking brandy, nor do I intend to. It might give a very false taste to the Claret.*] Then take your decanter in your left hand and your bottle in your right, hold the bottle so that the shoulder is between you and an electric light bulb, and pour slowly, gently *and continuously* until you see the little cloud of deposit commence to approach the shoulder of the bottle. Then stop; set your decanter in the dining-room with the stopper out; you should have done this about an hour before dinner, unless the wine is very old, and indeed I think there are few wines that will not stand an hour's breathing; young wines can stand more. Old wine sometimes has what is called 'bottle-stink' for a few minutes; take a second decanter, and pour from one to the other and back. This calls for no precautions, and the air reaches all portions of the wine more quickly; and you will find that the unpleasant odour is gone. But not if the wine is corked. Pour all night long, and you will never get rid of cork. Another bottle is the only remedy, not always available.

You have now rendered your penultimate service to your wine; it only remains to serve it in thin, white, large glasses, preferably of a tulip shape, and only three-quarter fill the glass. And all your trouble will have been wasted if you have invited the wrong people to drink it with you!

Most of these remarks would apply to other red still wines as well as to Claret; but because this wine is the most natural of all I have taken it as the arch-type; and it is the wine that will best help the student who is anxious to master his subject. Apart from the reasons already given, Claret is the easiest wine to study, because of the official classification of its wines made in 1855, which gives a kind of yard measure whereby to appraise and value its multitude of labels. There had been earlier classifications and the great wines had shown constancy in their pride of place; but the Chamber of Commerce of Bordeaux felt that the approach of an International Exhibition was a suitable occasion for revising the list, and they did it so thoroughly and well that no further classification has been deemed necessary since that day, although some of the classed vineyards might well alter their places in the list if it were to be revised in the light of recent history. There have been a few alterations in names, but here is the list as it appears in the 1929 edition of Cocks et Feret:

CLASSED WINES OF THE MÉDOC

First Growths

Château Lafite	Pauillac
,, Margaux	Margaux
,, Latour	Pauillac
,, Haut Brion	Pessac (Graves)

Second Growths

Château Mouton Rothschild . . .	Pauillac
,, Rausan Ségla	Margaux
,, Rauzan Gassies	,,
,, Léoville Las-Cases . . .	St. Julien
,, Léoville Poyferré	,,
,, Léoville Barton	,,
,, Durfort Vivens	Margaux
,, Lascombes	,,
,, Gruaud Larose Sarget . . .	St. Julien
,, Gruaud Larose Faure . . .	,,
,, Brane Cantenac	Cantenac
,, Pichon Longueville . . .	Pauillac
,, Pichon Longueville Lalande . .	,,
,, Ducru Beaucaillou	St. Julien
,, Cos d'Estournel	St. Estèphe
,, Montrose	,,

Third Growths

Château	Kirwan	Cantenac
„	d'Issan	„
„	Lagrange	St. Julien
„	Langoa	„
„	Giscours	Labarde
„	Malescot-St-Exupéry . . .	Margaux
„	Brown Cantenac	Cantenac
„	Palmer	„
„	La Lagune	Ludon
„	Desmirail	Margaux
„	Calon Ségur	St. Estèphe
„	Ferrière	Margaux
„	Marquis d'Alesme Becker . .	„
„	Boyd Cantenac	„

Fourth Growths

Château	Saint Pierre Sevaistre . . .	St. Julien
„	Saint Pierre Bontemps . . .	„
„	Branaire Ducru	„
„	Talbot	„
„	Duhart Milon	Pauillac
„	Poujet	Cantenac
„	La Tour Carnet	St. Laurent
„	Rochet	St. Estèphe
„	Beychevelle	St. Julien
„	Le Prieuré	Cantenac
„	Marquis de Terme . . .	Margaux

Fifth Growths

Château	Pontet Canet	Pauillac
„	Batailley	„
„	Grand Puy Lacoste	„
„	Grand Puy Ducasse	„
„	Lynch Bages	„
„	Lynch Moussas	„
„	Dauzac	Labarde
„	Mouton d'Armailhacq . . .	Pauillac

Château Le Tertre	. . . . .	Arsac
„ Haut Bages Libéral	. . .	Pauillac
„ Pédesclaux	. . . . .	„
„ Belgrave	. . . . .	St. Laurent
„ Camensac	. . . . .	„
„ Cos Labory	. . . . .	St. Estèphe
„ Clerc Milon	. . . . .	Pauillac
„ Croizet Bages	. . . . .	„
„ Cantemerle	. . . . .	Macau

There figure in this list sixty-two vineyards, in five categories, all of which would be called first-class wines. Behind these classed growths would come hundreds of vineyards classed as bourgeois, but which would do credit to any host; behind these again would be the *paysan* and *artisan* vineyards, some of which might furnish a very good wine, but most of which would normally be content to supply the market with its *ordinaires*. The classification was made upon the excellent economic principle of the laws of upply and demand, whereby prices are governed by the demand for the article. It was found that the three great Médocs, Lafite, Margaux and Latour, consistently outdistanced all rivals (with two exceptions) by such a margin as to make them a little group on their own. The two exceptions were Château Haut Brion, a red Graves wine, and Château Ausone, a Saint Emilion. These were both voted the peers of the three Médocs, and Château Haut Brion was actually brought into the classification; this compliment was only withheld from Château Ausone because of its small output—15 tonneaux against 100 at Haut Brion, 180 at Lafite, 100 at Latour and 150 at Château Margaux. [CAMPBELL: *My 'Cocks and Feret' of* 1886 *gives the average production of Château Ausone as* 8 *tonneaux, not even* 15, *but says nothing about the withheld compliment: I have heard rumours of it in recent years, and would be glad to have confirmation.*] Following the first group there was (in those days) a well-marked gap in the prices before coming to a second group of sixteen vineyards, which again closely jostled one another in the list. The next groups were equally well marked, and numbered fourteen, eleven and seventeen respectively; then there was a considerable drop before the bourgeois wines made their appearance. But if the classification were to be made to-day there would be some remarkable shifting of places. For the moment (but, I hope, only for the moment) Haut Brion would go out of the list. The vineyard changed hands between the vintages of 1922 and 1923. Monsieur Gibert, the new proprietor, was a stranger to viniculture; and although his first wine, the 1923, was one of the most beautiful of that very charming vintage, and his second, the 1924, was fairly good, the succeeding Haut Brions fell very far beneath the high standard expected from that Château. The 1926, 1928 and 1929 are great, heavy, coarse and ropy wines, lacking all pretence to breed. I have not tasted any of the more recent vintages; but the Château has now passed into the hands of Mr. Dillon, an enthusiastic wine-lover from America, who is determined to restore the ancient pre-eminence of this noble estate; and nobody will rejoice more than I in his success. I venture to prophesy great things for Haut Brion 1937! [CAMPBELL: *Not yet fulfilled, but the whole vintage is rather backward.*]

But if Haut Brion must be momentarily omitted from the first growths, what about Château Margaux? The last really good Château Margaux that I have tasted was in 1920; the 1924 was a pretty wine, but not up to the standard of a first growth in a good year. In 1929 they allowed me to taste their wine of 1927; I had spent a morning spitting out the '27's of other vineyards, but the Château Margaux 1927 was smooth and sweet, and methought the lady did protest too much. However, my one meeting with the 1929 led me to believe that any momentary falling from grace had been redeemed; and I shall hope to find the '33's and '34's, and in due course the '37's reminiscent of the great days of Château Margaux.

If Haut Brion's place be for the moment vacant there can be no doubt about the claimant. Under the ownership of Baron Philippe de Rothschild, Château Mouton has steadily improved, and for many years has commanded prices higher than any of the first growths. And I owe Mouton an apology. The 1920 Bordeaux vintage was trumpeted at the time as likely to be one of the outstanding vintages of the century; and we patiently awaited their development, I confidently expecting to drink them with delight in 1935, or thereabouts, for they were full of tannin and slow to come round. But to my horror about 1930 I began to see a presage of wear in them; I had bought some Mouton 1920, and I began to drink it, unwillingly but hopefully. Alas! not a single bottle had yet ripened and still each tasted as if it had gone over the edge. So that I, writing a little book on Claret in 1934, sang the swan-song of my Mouton, and indeed of all 1920's, lamenting them as Mary and Martha may have wept for Lazarus. For lo and behold! the 1920's were not dead; they were sleeping. Last spring Robert Speaight gave me dinner one night at the Savile Club; and he crowned the feast with a Claret that summed up all the glories of the wine. It was the 1920 Mouton. I thought of Browning's lovely image at the end of 'One Word More':

> " But the best is when I glide from out them,
> Cross a step or two of dubious twilight,
> Come out on the other side, the novel
> Silent silver lights and darks undreamed of,
> When I hush and bless myself with silence."

For the wine had 'come out on the other side,' and was at last showing the beauties so long hidden; and the only fit deed was to 'hush and bless myself with silence.' So, my dear Robert, let me thank you again; and make your beautiful voice my intermediary to negotiate my pardon from the lips of Mouton!

The 1920's seem all to have come back; and in the light of my experience with them and my recent experiments with the 1924's, I should be inclined to advise my friends to lay by their '24's for a year or two, as I think they are undergoing the same curious development. They have certainly lots of life in them yet; but they are not drinking satisfactorily in 1940. I have every hope that they are only undergoing the same translation as the 1920's have undergone, and that in a few years they will be even more beautiful than they were a few years ago.

Coming back to our classification, I confess that I have a very great *tendre* for Léoville Poyferré, the 1914 wine of which vineyard will always have precious memories for me; for it was the first classed wine that I selected for myself. It was very lovely, and stood up very well to its elder brother the 1900, which was a superb wine, though at times I think the younger wine beat it. So that I am not sure that I should not lift the Poyferré, if not actually into the first growths, at any rate to the head of the seconds. In the same way I should be inclined to promote Giscours from the thirds to the seconds; it is an admirable wine, and its 1920 was second only to Latour. I have not tasted this since its 'resurrection,' which I hope has taken place. The list of fourths includes two wines that have never given me a bad bottle: La Tour Carnet and Beychevelle; I always think of Beychevelle as a 'second,' and La Tour Carnet can hold its own in any company. But how Pontet Canet has been content to remain a fifth growth I cannot understand. I have always found this wine elegant and full of breed; and in 'blind' guessing I have invariably named it as a second, if not, as I did once or twice, a first. You may think this a prejudice on my part in favour of Cruse wines; but I have no such *tendre* for Rausan Ségla, excellent though that wine is, and I prized Pontet Canet long before I first visited Bordeaux and became for ever a debtor to Messieurs Christian and Emanuel Cruse. I would also promote the two Grand Puys, which have always seemed to me most deserving wines, with the true gentlemanliness of Pauillac; they have served me well, and left none but pleasant memories behind them.

I have only once tasted Château Boyd Cantenac, which suddenly reappeared in the classed list after an absence of forty years, which occurred in peculiar circumstances. The vineyards of Boyd, Brown and Abel-Laurent are intermingled in a most confusing way; and it so happened that Monsieur Abel Laurent, a wealthy banker in Paris, having made a name for the wine of his own vineyard, which was classed as a bourgeois, bought the Boyd vineyard, and found it practical to sell both wines under the same familiar label. Château Boyd then fell out of the classed list. Twenty years ago, when the Boyd vineyard came under separate ownership again, the new owners made a claim to be restored to the list, and there they are. I say 'they,' but the estate is now owned by Monsieur F. Ginestet, the well-known shipper (and, incidentally, one of the best after-dinner speakers I have ever heard).

I am told that the two Gruaud Larose vineyards are now under one ownership again, as are the two Saint Pierres.

But enough of the lists; let us have a look at some of the wines and see what we can remember about them. That kindliest of critics, Ian Campbell, suggested that my last book might have been more useful if I had said something about the unclassed wines; so if I become garrulous, let him share the blame.

CLARET (II)

I WONDER when I began to take notice of Claret labels? The fact that my father was a member of Parliament brought to our letter-box a large number of circulars, a Mr. E. Godfrey Felton being particularly eager to satisfy our possible needs. Pink and blue and yellow and red lists used to proclaim the most wonderful bargains in wines; it must have been in their pages that I first learned that Château Lafite and Château Larose were notable Clarets, and Château Yquem a wonderful and very valuable wine. The fact is eloquent of the primacy obtained by Lafite, and of the imposition upon the public of the label 'Château Larose' by the lazy Thackeray, who when he wrote of Château Larose was referring to the great vineyard of Château Gruaud Larose Faure Bethmann [CAMPBELL: *The Faure Bethmann was never divided but was itself a division of the Château Gruaud Larose*], perhaps not then yet divided into two, now again one vineyard. 'Château Larose' *simpliciter* is a label to be regarded with the gravest suspicion, as there is no classed one so called, although there is a Château Larose in the Entre-deux-Mers district (which sometimes spells its name with a Z), and there is also a Château Larose in Baurech on the right bank of the river, opposite Graves, which produces a white wine of some fame. But a Claret sold as Château Larose is an inferior wine, not to be introduced to the society of the Gruaud Laroses and their peers.

The primacy of Lafite amongst Clarets has been responsible for a great deal of '*snobisme*'; hosts are afraid to offer any wine of a lower order. Without in the least detracting from the greatness of Lafite, I desire to protest against this absurd error. To give a simple example, a kind friend recently sent me a case of the Lafite 1933 and another of the Gruaud Larose Faure 1928. I have frequently served them at the same meal, and the second growth has always been pronounced much better than the first. Of course, 1933 is a light vintage and 1928 is a very big wine indeed, so that the years are in competition as well; but it is by no means unusual to find in a particular year a wine of inferior rank in the hierarchy overtopping its superiors. Take, for example, 1920. I always thought the Château Latour far and away the best wine of that year; but I should put Château Giscours second, although it is ranked only as a third growth; and I think I should put Léoville Poyferré third, although a second growth, and Lafite fourth, although a first. [CAMPBELL: *As Maurice himself would say, 'de gustibus non est disputandum!' and I wonder how often he and I and other good friends violated the excellent precept.*] The Château Margaux 1920 did not seem to me amongst the best, and the Haut Brion 1920 has proved a very variable wine in my experience. But Château Laujac, a little bourgeois wine of almost invariable goodness, surpassed itself that year, and the 1920 Laujac outshone several of the third and fourth growths. I was delighted to come across it once more a few months ago, and although it had worn a little thin it was still a delicate and beautiful wine.

But Lafite has filled my glass with wine and my heart with gratitude probably more often than any other wine. My acquaintance with its excellence extends over the remarkable stretch of 128 years! for in 1926 it was my privilege to be present at the Château when over three hundred guests were given the wine of 1811, the year of the Comet! And at its age of 115 years it still drank graciously, with not more than a suspicion of fading. I think it must have been the greatest Claret ever made. [CAMPBELL: *I never tasted the* 1811 *Château Lafite which no doubt must have been a remarkable wine. Maurice had the good luck to drink it at the Château where it had reposed peacefully for more than* 110 *years; also it was no doubt served at the end of a sequence of wines of the same vineyard and of increasing age. His admiration can be well understood and his ecstatic judgment received without reserve. Château Lafite of* 1864 *is the finest Claret I have ever drunk and I am satisfied.*] Mr. Warner Allen makes this claim for the 1864, basing his opinion upon the reported experience of somebody who had tasted the 1811 in 1868 and must have been very unlucky in his bottle, for it was vapid and tasteless; whereas Mr. Warner Allen in 1921 found the 1864 full of vigour, as it continued to be for years afterwards. Well, I found the 1811 magnificent in 1926; and of all the party there was not one critic to suggest that the wine was *passé*. And, although the 1864 is only 75 years old, I have found my recent experiences of it less satisfactory than that of the 1811. I have had the luck to drink the 1864 in many forms: bottles, magnums, jeroboams and imperiales. To say one word that would seem to detract from its magnificence would be unjust. But I fear it has at last reached the limit of its long and honourable life; and the fact that it has not equalled the 1811 in longevity does not seem to me to be a serious demerit. And I doubt that there would be very much left to attain that age; its merits were better appreciated than those of the older wine, and it must be many years since it was possible to purchase 1864 Lafite.

I am not sure if it is still obtainable at the Trocadero; but some years ago through the kindness of Ronald Avery of Bristol I was enabled to drink it there against the Château Latour of the same year. The Lafite romped home; the Latour had not stayed the course. But if there be rivalry between these great wines, it is only fair to say that Latour won equally handsomely in 1878; the imperiale of that vintage which we drank at the Saintsbury Club in October, 1931, was one of the greatest wines I ever tasted [CAMPBELL: *Hilaire Belloc expressed a similar opinion on that memorable occasion*], and an imperiale of Lafite of the same vintage which we drank in 1932 was very disappointing after our happy experience with the Latour.

I have drunk the 1847 Lafite; then come the series of classic vintages, in all of which, I think, I have tasted it: 1858, 1864, 1865, 1868, 1869, 1870, 1871, 1874, 1875 and 1878. I think I then jump to 1889, 1890, 1893, 1895, 1896, and so to the beautiful 1899, one of the most exquisite Clarets ever made. The 1900 was also very fine at Lafite, but I should not reckon the Lafite 1900 amongst the best of the year; both Château Margaux and Château Latour were better, I think, and the Mouton certainly was, although all of these had been uncommonly good in 1899 as well. I never came across the Lafite 1904, but the 1906 was a gracious wine, and the 1909 was very good indeed. My friends know my eccentric partiality to 1912, but the Lafite was not amongst my favourites in that misprised year. Oddly enough, it figured on the dining-car list of the London, Midland and Scottish Railway Company for some years: not, one would imagine, the most favourable place in which to experiment with an unfamiliar wine. The 1914 was again an exquisite wine which, however, held even less well than other wines of that strange year. The 1920 was very good; the 1921 was a failure and not put on the market. I once got some 1922 cheap and drank it as an *ordinaire;* it was a very good bargain as such, although hardly to be called a good wine. The 1924 was easily the best of the year; it is a wine that cannot be praised too highly. I think that the 1933 is the only wine I have tasted since, and it is light and elegant.

Oddly enough, although I esteem Latour more highly even than Lafite, I have not drunk it in at all the same range or variety. I used to disagree with Professor Saintsbury in his setting of this wine first of Clarets; but its unvarying excellence gradually converted me. For example, I have spoken above of the Lafite 1922 as an *ordinaire;* but the Latour 1922, without reaching to its own standard of greatness, was a really delightful wine, which would have deceived the elect into a guess of something much greater. The 1878 of which I have already spoken will always be my greatest memory of Château Latour; but I once possessed a little of the 1879 wine, which was entirely admirable, and added to my reputation for choosing unorthodox vintages [CAMPBELL: *He himself has said it! he was also very obstinate, and if, as he says two lines lower down, his* 1879 *Latour won him many compliments, it shows what his friends were prepared to expect from his unorthodoxy!*]: for 1878 is usually considered the last of the prephylloxera vintages. But my 1879 Latour won me many compliments; it was entirely free from disease, and although a trifle on the weak side, was a well-balanced wine, of breed and finesse. Of the 1912's, Latour was easily the best. I remember once drinking it against a La Tache 1904; we had intended to use the Claret as a mere preface to the Burgundy, for anyone who had the good fortune to taste La Tache 1904 at its best will seek no assurance that it was in need of no courtiers. But by an accident our Claret was removed and successfully recovered; and we went back to the Latour after the La Tache, and the younger wine suffered nothing by the new comparison. This is probably the most severe test to which a Bordeaux wine could be subjected; delicacy is not always easy to appreciate after the robust vigour of a more soldierly wine. Again, in 1920, Latour overtopped all competitors. I remember on my first visit to Bordeaux in 1924 asking for the best wine that the *sommelier* of the Hôtel de Bayonne could produce for me. When it came I said reproachfully: 'But this cannot be older than 1909.' It was the Latour 1920, already in magnificent condition; and I have never drunk a bad bottle of it since. The 1924 was about its least notable success; it may be that even yet the Latour has not arrived, for it is a wine to keep (although the 1920 seems to have contradicted this by its precocity; but then wines mature more quickly in Bordeaux than in England). I have not tasted any later vintage, I think.

With Château Margaux my acquaintance has been even less extensive. I have tasted the 1870, but I then pass straight to the 1893, and then to the 1899 and the wonderful 1900, best wine of that very lovely year. I think I have only cellared three vintages of this great Château: 1906, 1907 and 1924, all in very small lots. The 1906 was the only satisfactory wine of that year that I have met. I had it in magnums; and I have not tasted it for a very long time, so that it may have come round. The 1907, so long as it lasted, was a very agreeable wine indeed. They had a good deal of it at Gray's Inn, where they used it as their Claret of ceremony as long as it lasted. I think they also had it at the New University Club. At any rate, I had a fairly wide experience of it, and I thought it a wine that did credit to Bordeaux. The later years did not often come my way; I have drunk the 1912, 1914, 1915 (yes!), 1917, 1919, 1920, 1923 and 1924. None of them were really noteworthy except the 1920, which was a good wine, but not, I think, amongst the first half-dozen of that year. But the wine that is bottled at the Château under the label 'Le Pavillon Château Margaux' is worthy of a wider appreciation than it enjoys. I am not sure that I am right in thinking that originally the Pavillon was a separate vineyard; now it signifies a second wine of the principal estate. But I never enjoyed a light Claret so much as the Pavillon Margaux of 1912. It had everything except weight that goes to the making of a delightful wine; you felt you could go on drinking it indefinitely and with undiminished pleasure. It was very like a light edition of the Château Haut Brion of that year, a year that was badly afflicted with mildew; but I suspect that I must like mildew, for I never drank a 1912 that I did not enjoy. [CAMPBELL: *Maurice would not himself admit that the* 1912 *vintage was a victim of mildew, and I think he here has a sly dig at myself who always maintained that it was.*] The only other vintages of the Pavillon that I have come across were the 1914, which was very good, and the 1916, for which I did not care very much. I think that the 1916 was the first wine labelled on a qualitative basis rather than a geographical.

Of the second growths, my first favourite has always been Léoville Poyferré, although I think I always mentally class Mouton as a first and not a second. With Mouton I have got back as far as 1865, and on two occasions I have drunk the '70, wondering whether it would ever reach its full development, for even in recent years there still seemed to be something unsaid in its depths. The 1899 and 1900 were the best Moutons I had the good fortune to meet. At the Château they had a very curious belief in their wine of 1916, which seemed to me unwarranted, as I thought it rather harsh and dry. I have hardly tasted the later vintages, but in 1950, if I am spared so long, I shall hope to try the 1929, which should be about at its best by then. [CAMPBELL: *Sad that he will not be alive to celebrate the very probable fulfilment of his prophecy.*] Now, I never feel the same awe when dealing with the Léovilles as I do with Mouton; perhaps the wine of St. Julien is more friendly than that of Pauillac. Certainly, I have always felt at my ease with Léoville Poyferré; and between its 1914 and its 1900 I must have drunk as much of it as of the produce of any other vineyard. I have had a few old bottles, and some younger ones; the 1920 was very good, and the 1924 was probably second only to Lafite, while the 1929 was very full and gracious. I have a surprisingly small acquaintance with Léoville Barton, and of the Lascases wine my principal memory is the lovely 1906, oddly enough one of the most elegant and delicate wines I have ever tasted, although 1906 is the last vintage to which one would expect those adjectives to be applied. Although I have pleasant memories of almost all the other second growths, none of them are outstanding favourites of mine. Amongst the thirds I put Langoa first, largely, however, because of my pleasant recollections of its hospitality. But I have had some lovely Langoas, in particular the 1900; and the 1914 was an uncommonly good wine. A third growth that I used to love for its silkiness has, alas! ceased to be. Château d'Issan could not stand the misfortune of three successive bad vintages, and in 1932 ceased to exist. In 1929 I was told that Château Ferrière had ceased to produce a wine, but I do not think this was correct, although I have seen no recent vintage of this estate. The beautiful Giscours cut down the acreage of its vineyard, but will, I hope, long continue to give us its delightful wine. And if d'Issan has gone, Boyd has returned to take its place.

Beychevelle and La Tour Carnet always seem to me to be miles away the best of the fourth growths; but I confess I have found some difficulty in coming across many of them. For some reason, there was a sudden spate of Château Talbot 1929 on the market recently, and at a very modest price, for which I think there must have been some accidental explanation, for the wine drank well. I have heard that the installation of glass-lined fermenting vats at Château Talbot has not been as successful as had been hoped. The two Saint Pierres have been reunited into one vineyard; they both were shy birds, and I think I have had only a couple of meetings with each. Recently in a northern list I found a lot of Château Rochet 1911, which proved a very pleasant bargain. Up to that I had never met this wine.

Finally, what on earth is Pontet Canet doing down amongst the fifth growths? Here is a wine which is never common placeand almost always distinguished; it always seems to me to have much of the breed of its near neighbour, Lafite; and it is the least temperamental of Clarets, always conscious of its duty to please and refresh and never failing to do so. That my opinion for once coincides with that of the orthodox is evident from a consideration of the price the wine always commands; it competes with the second growths. The 1929, to which I gave my personal benediction (by sleeping comfortably in the Château while the wine was being made), is about as fine a specimen of Bordeaux red wine in a good year as I have ever drunk. The 1928 has been, like all '28's, a laggard; but it is at last decking itself for the party, and putting on a countenance that promises great delight to its company in the immediate future.

And let us do justice to other fifth growths, junior only to the super-excellent. The Grand Puys have never given me a bad bottle; some of my adventures with this vineyard (for it once was one) have been amongst the greater moments. The Lynches have proved themselves agreeable; but they have a curious Irish quality of never showing themselves at their best in company not of their own choosing, and I doubt that they would have chosen me. You see, Count Lynch was the implacable enemy of the gallant Marshal Ney, and hallooed on the hounds that were hunting him to his death. The man that called Ney a traitor would not feel comfortable in my company. [CAMPBELL: *I have drunk Lynch Bages so good as to be capable of reconciling the two Irishmen.*]

Dauzac is another wine that I think has been classed much too low. I have never drunk a bad bottle of Dauzac; and the 1920, 1924 and 1926 vintages gave me some very pleasant drinking. I should have imagined Dauzac worthy of promotion to the third growths. But I must not use up all my favourable adjectives on Dauzac, for here is Mouton d'Armailhacq waiting for my comments. I believe that Baron Philippe de Rothschild has now united the vineyard to the great Mouton, the wines remaining separate, but the care and attention being the same. No greater compliment could have been paid to the junior wine, but it fully deserved it all the same. For Pauillac has never been more agreeable than when poured from a bottle with the d'Armailhacq label. I do not wish to be misunderstood; I am not saying that it is the peer of Lafite or Mouton. But after all, price does mean something; and you can drink a fifth growth at little more than half the expense attaching to a first or star second. And, at its modest price, it has all the elegance of the greater wines, although perhaps not quite the same body. I have so many delightful memories of this charming wine that I wonder how much of the praise belongs to the wine and how much to the friends with whom I drank it. For to-day I must always turn down an empty glass for poor Alan Freeman, with whom I shared so many good bottles, little suspecting how soon a tragic accident was to interrupt that good fellowship. *Requiescat!*

Clerc Milon and Cos Labory share the triumph of having so far completely eluded my pursuit; they are the only classed wines of which this remains true. Some of the others have only afforded me a very inadequate tasting, for example, Cantemerle, which is supposed to be a Dutch favourite, and which pleased me on the only occasion on which I met it. I once stumbled on a lot of Camensac 1924 in Imperial Pints, and took the lot; it made very good drinking while it lasted. There is much to be said for the Imperial Pint as a Claret bottle; a half-bottle is sometimes a little niggardly for one person, and a bottle is almost always too much if you are alone. The Imperial Pint does it just nicely.

This deals with the classed wines of the Médoc; but there are other labels calling for a comment. I do not think I have ever come across the Château Bel Air-Marquis d'Aligre, which straddles across the boundaries of Soussans and Margaux, and the wine of which has always commanded a price comparable to those of the classed wines. Another of these half-bloods is Château Chasse-Spleen, which is a Moulis; I have drunk many agreeable bottles of this, although Harry Huggett could not see any merit in it. The Inner Temple used to serve it in Hall, but it was before my acquaintance with the wine-list, so that I cannot be sure of the vintage; it was probably a poor year if it gave dissatisfaction. I have never come across a wine singled out by the editors of *Claret and Sauternes* for citation as exceptionally good—Château Pomys, a Saint Estèphe which is classified as 'premier bourgeois,' but which, I am told, will often overtop wines classified above it. The same I can myself assert of its neighbour, Château Phélan-Ségur, a name where Irish and French blood mingle, as they did so often on the battlefield. I have already spoken of the excellence of Château Angludet, which has never failed to please me, although sold for a very moderate price; and I must also give grateful thanks to Château Citran for many a satisfying glass, not beyond the range of wartime economy. For those who like their wine to be flavoured with history it may be mentioned that Château Citran, for many years the property of the Clauzel family, was formerly owned by a famous Catholic house, the de Donnissans, the last of whom died for his King in company with his cousin, Henri de Rochejaquelin, in the Vendean rising of 1793. His daughter, widowed in the war, subsequently married Louis de Rochejaquelin, and they lived at Citran until 1832, when they sold the property to the Clauzels. The wine is as loyal as its former owner, and the vintage of 1929 is worthy of a connoisseur's attention. A neighbouring vineyard (they are both in the commune of Avensan) is Château Villegorge, classified as a Grand Cru Exceptionnel; its output is only 20 tonneaux of red wine per annum, which may explain why I have never come across it. Smallness of output has prevented many notable vineyards from appearing in the classified list; Château Ausone is probably the most notable example, but in Margaux there is the Château Dubignon-Talbot, which was formerly classed as a third growth, but which only produces 8 tonneaux a year, and has thus been de-classed, although the wine remains excellent.

There is only time to glance at one or two less well-known estates whose wines I have found agreeable. In the parish of Parempuyre there is a Château of that name, owned by Monsieur Fréderic Crusé, where a wine is produced that maintains the high standards to be found in all the vineyards associated with that family. In Le Pian is found Château Sénéjac, the 1933 wine of which has raised many hearts at the Inner Temple in recent times; I have not often come across it. In Margaux are two neighbouring estates with a punning resemblance of names, one called Château de Labegorce and the other Château de l'Abbé Gorsse-de-Gorsse. I have found the wine of the former very agreeable, but I have never come across the latter, although it has a high reputation. Half the wines of Moulis appear to be sold under the label 'Au Grand Poujeaux,' no fewer than eleven proprietors so designating their wines. As some are mere *artisan* and *paysan* growths it is wise to look for a supplementary description, inserted by the five owners of bourgeois estates. Château Listrac in the commune of that name produces some outstanding wine, the 1929 being a splendid example of how good a Claret can be without being expensive. And, speaking of good, cheap wines, I should like to say a word of praise of Château Loudenne, an estate owned by Messieurs W. & A. Gilbey, and lovingly fostered by them to maintain the firm's standard Claret and Graves. These wines are sold at a recognized modest price, and some people imagine that good wine cannot be made to sell at such a price. Two experiments can correct this error: a tasting of the wine and a visit to the vineyard. Messieurs Gilbey have the resources of a very large distributing business behind them; and they have spared no expense in developing their estate in Saint Yzans, north of Saint Estèphe. The vineyard is a very fine one, and the entire establishment is run on the most up-to-date lines. When I last visited it they had just put in glass-lined vats in which to work their fermentation. I should like some information as to the success of this experiment, for other vineyards have tried a similar experiment, and at some it has not been very successful; apparently the must needs to breathe through the wooden tonneau. I was also interested to find an attempt being made to produce a French equivalent to Port, which was to have been put on the market under the name of 'Medulio'; but I have never since heard of this, and I fancy that the experiment was abandoned. It may gratify Messieurs Gilbey to know that when I shared my mother's house, no matter what I offered to my guests or drank myself, there was always a bottle of Château Loudenne for her own consumption. And I did not despise a glass of it myself.

But we have now gone as far to the northward as will be profitable (for I have already made reference to Château Laujac); let us make our way back to Bordeaux, preparatory to another excursion in the opposite direction.

CLARET (III)

WHEN the Médoc was uncultivated land, unsuspecting its future glory as the little girl never dreams of the lustre that her great-grandchildren may bring to her own name, the district now known as Graves was the true home of Claret. Henry II linked his French and English dominions; a generation later his son John was already importing red wine from the Gironde, and the best of it came from the vineyards just south of Bordeaux. When the wind blew from the sea, carrying with it the tang of salt water, it must have picked up the smoke of the city and deposited its finer ash on some of the greatest vineyards of the world. Bordeaux has grown; to-day it can almost touch the most famous of all the estates whose wine has helped to make the city's glory. A penny tram-ride along the road leading to Arcachon brings the traveller to the gates of the world-renowned Château Haut Brion, with a wine so excellent that the authorities engaged in classifying the great growths of the Médoc in 1855 felt that they would stultify themselves if they did not give the freedom of their list to this estate beyond the frontier, whose wine had been for generations the rival of Lafite, Margaux and Latour. And the English visitor, especially the Londoner, should cross the threshold of Haut Brion with more than ordinary interest, for it is probable that more of the wine of that vineyard has gone down London's throat than of any other estate in the world. My reason for saying so? Listen.

In 1691 Richard Ames published *A Search after Claret*, in which he bewailed the impossibility of obtaining a glass of that wine in the city of London. William III had ascended the throne; war with France was in future to be conducted upon an economic as well as a martial basis; and for the first time since the days of Henry II it was impossible to import the wines of Bordeaux. Ames imagines a tour of the taverns of London, where it is possible to get red or white Port, Hock, Canary, Florence or Sherry, but not a drop of Claret. The point to which I wish to draw attention is that Ames's way of referring to Claret is 'a glass of Puntack.' Now, Puntack or Pontack took its name from the famous eating-house, Pontac's, which was founded by the son of Evelyn's 'famous and wise Prime President of Bordeaux,' whom he describes as the owner of the Château Haut Brion and Pontac. Young Pontac had come over to London shortly after the Great Fire and had set up an eating-house on the site of an old tavern called the White Bear, which stood at the corner of Lombard Street and Abchurch Lane. Why should a sprig of the French nobility come over to London and start a restaurant? If he had already a big *clientèle* for his wines one could understand it. Had he? Let us go back a few years to April 10th, 1663, when we find Samuel Pepys going to the Royal Oak Tavern in Lombard Street, where he 'drank a sort of French wine called Ho Bryan that hath a good and most particular taste that I never met with.' Now, in 1663 the name 'The Royal Oak' was three years old; one of the manifestations of joy at the restoration of Charles II. What was its name before? We know that as far back as 1365 there was a house in Lombard Street for the sale of French wines, one of three in the whole city at that date. Is it leaving too big a gap in the argument to suggest, first, that the White Bear was either that house itself or its successor in title and goodwill, and second, that it was the White Bear that changed its name to the 'Royal Oak'? When the Royal Oak was burned down a couple of years later, there would still be many who would speak of it under its older name; and I find it difficult to believe that the owner of a French vineyard should sell his wine to two rival houses in the same street. That Pontac set up a restaurant is certain; that he would have chosen the site where his wines were known is probable; that he chose the site of the White Bear is traditional; *ergo*, the White Bear and the Royal Oak were the same house. Not a watertight syllogism; but one that has probability on its side.

Pontac's success as a *restaurateur* is proverbial; but we have the evidence of Richard Ames that 'a glass of Puntack' was synonymous with the best Claret. The Château Pontac stood where the Château Dulamon now stands; but in those days its output was white wine. The red wine at Pontac's was that of his other estate, Château Haut Brion, 'that hath a good and most particular taste.' And therefore I suspect that London learned to judge all Claret by the standard of Haut Brion, which will explain why English Claret is esteemed the best in the world, now that its importation into Ireland has virtually ceased. I have given my reasons elsewhere for my theory that Château Haut Brion originally was Château O'Brien; and the more I consider them, the more do I think they may possibly be correct. [CAMPBELL: *Nobody took Maurice's claim seriously except himself, and on his part I think it was a bit of intentionally humorous make-believe.*]

Whatever the origin of the name, Château Haut Brion has had an illustrious history. It was owned for a few years by Talleyrand; he sold it to a bank, who in turn sold it to a Monsieur Larrieu in 1837, and it remained in the Larrieu family until 1922, when it was purchased by Monsieur André Gibert. He in his turn sold it a few years ago to an American gentleman named Dillon. It is to be regretted that during Monsieur Gibert's ownership certain experiments undertaken by him did not at all improve the wine. The last really notable Haut Brion that has made its bow is the 1923; the 1924, though fairly good, is not an outstanding example of what the great vineyard can do: 1925 is not a year in which to pronounce on any Claret, so that it is not until 1926 that we get a fair chance of judging how the change of ownership has affected the output of the Château. And I have given a fair trial to the Haut Brions of 1926, 1928 and 1929; and I pronounce them to be hot and unpleasant and entirely unworthy of the name of Claret. [CAMPBELL: *I agree. ''Tis true, 'tis pity; and pity 'tis 'tis true.' Happily Haut Brion is staging a come-back: it was a marvellous wine in the past and can be again.*] I have no idea how they were prepared; but if I were told that some attempt had been made to strengthen the wine by reducing some of it and adding the boiled product to the vat I should then expect the taste to be exactly what it is. The Bordelais sneer at Burgundy as a cooked wine, because, they say, the Bourguignon helps his fermentation by boiling some of the must and returning it to the fermenting vat. I have often tasted self-styled Burgundies that suggest such a course of conduct; hot and heavily perfumed, they have lost all the delicacy of a natural wine without acquiring the agreeable qualities of a Port. It is precisely this thick, hot flavour that I get from the Haut Brions of which I have spoken; and I cannot imagine any of them turning into drinkable Claret in any number of years.

Fortunately, Mr. Dillon has entered into his new estate with a firm determination to employ none but the best traditional methods of making wine. Only the grape vines that can show the noblest pedigree are to be planted in replacement of the exhausted roots; I believe they are to be more widely spaced, so as to give each vine a larger area from which to draw, and the sacrifice of quantity thus made will probably be more than repaid by an improvement of quality. Nature is to be given a free hand in the making of the wine, and the innovations of recent years have been repealed. In ten years' time I hope to drink the Haut Brion of 1937, the first reasonably good year in Bordeaux since Mr. Dillon took over the Château. And if, as I hope and believe, his wine proves then and thereafter a noble vindication of the old tradition of Bordeaux, he will have no more delighted or grateful admirer than myself.

I have spoken elsewhere of the excellence of the white wine made at the Château; I have not tasted any recent vintages of this, and do not know if it also passed under a cloud, but I trust that the loving care now lavished on its red sister will not prevent due attention being paid to this most admirable of white Graves.

While there can be no doubt about the primacy of Château Haut Brion amongst the vineyards of Graves, the nobility around the throne are formidable in number and magnificent in quality. There would be keen rivalry for the viceroyalty; personally, I should give it to Château Pape Clement, although many would probably vote for Château La Mission Haut Brion, and not a few would suggest the Domaine de Chevalier. I have a special *tendre* for Pape Clement; when I first joined the Munster Circuit in 1910 and undertook the care of its wine, we still had some Pape Clement in Cork, which I suspect was the 1899 (for, alas! I was a teetotaller in those days and knew nothing of Châteaux or Vintages). It was highly esteemed by our *cognoscenti*, and drunk by the others. The vineyard has a more important link with the lawyers; it takes its name from Bertrand de Goth, Archbishop of Bordeaux, who once owned it, and must have regretted leaving it to take up his abode at Avignon when he was elected Pope, and took the title of Clement V. He had bequeathed the estate to his successor in the see of Bordeaux; and its history has continued honourable. But Clement V, living at Avignon under the protection of Philip le Bel, King of France, allowed himself to be persuaded by that monarch to suppress the Knights Templar, who had become a menace to more than one king in Europe. That they were a danger to the integrity of the French kingdom cannot be doubted, but this does not excuse the monstrous campaign against them, in the course of which every form of vilification was employed, and the noblest members of the Order were put to death in ignominy. In England, Edward II confiscated their estates and gave them to the Knights Hospitallers; after a short interval, unchronicled in any surviving record, we find the property of the Temple vested in the lawyers who have held it ever since. And yet we have no Château Pape Clement in the cellars of the Inner Temple, nor, so far as I am aware, is there any in the Middle Temple, either!

La Mission Haut Brion takes its name from a community of the order of Saint Vincent de Paul. The wine has always been famous, and in some years, notably in 1899, it has been a formidable rival to Haut Brion itself. I would rank both Pape Clement and La Mission with the second growths of the Médoc, although others lower them to third rank. I would more readily set there another delightful red Graves, Château Smith Haut Laffitte, a modest wine of great excellence that never asserts itself. I have drunk admirable bottles of it, the 1920 in Berlin and the 1924 here in London; and that 1924 was drunk in the company of my dear friend Edward Bunyard, who might have stood as a type of the wine in his excellence and modesty. Alas! that we should never renew our memories of that bottle together!

I have never been fortunate with my Château Haut Brion

Larrivet; the bottles were right, but only just right, like the Scotsman's change. The proprietors waged legal war with varying fortunes against the great Château, which sought to prevent them using the name 'Haut Brion' otherwise than as a mere addition; but the Cour de Cassation finally decided in their favour, and the Haut Brion now comes first in their title. Oddly enough, my best memory of this wine has to do with that poor year, 1925, when most Clarets were very inferior. But Haut Brion Larrivet was in that year lucky; its wine was fuller in body than most, and it had a delicate flavour. I had a couple of dozen, which served me very loyally. It is strange that another very good red Graves should have given me good aid in that doubtful year; I refer to Château Carbonnieux, a name more often associated with white wine. I have never found any special merit in the white Carbonnieux, but I am very fond of the red, which I have known in many vintages, notably the 1914. But the 1925 Carbonnieux was quite a friendly wine, and it made me a friend of the vineyard for life. I once picked up a very odd parcel, made up of three bottles of this vintage and two of that, some three dozen in all; it included the 1918, 1920, 1922, 1924, 1925 and 1926. It will be observed that there are three doubtful, two good and one fairly good years in that list. The odd thing is that I cannot remember the good years standing out above the mediocre! Perhaps the whole lot had been unfortunate in its cellarage.

At Château Haut Bailly Monsieur Malvezin developed a technique in making his wine which was borrowed from Pasteur. He claimed that by putting his wine through this process it aged much more rapidly, and attained maturity in a few years instead of the traditional dozen. I have drunk some very fine bottles of Haut Bailly, and have the most affectionate memories of the 1899, and only slightly less of the 1909, but I have not been fortunate with my later vintages here. I do not know the period covered by Monsieur Malvezin's experiments, so that I cannot pass any judgment upon the question whether he has been justified; but if you see a Château Haut Bailly of a good year on the wine-list do not hesitate to order it; you are certain not to get a bad bottle. [CAMPBELL: *This seems to me to be a very rash state-*

ment: nearly all clarets register an occasional failure. I cannot thole the 1929 *Haut Bailly, for example.*] When last I heard of the estate it had passed to new proprietors; I do not know if they have held to the Malvezin tradition or not.

These vineyards which I have named include practically all the red Graves one finds mentioned in English wine-lists. But adjacent to Haut Bailly is an estate that I commend to the notice of all Claret-lovers: Château Malartic Lagravière. My acquaintance with this wine is one of the many good things I owe to Barry Neame, who cellars the 1923, and I have said before that nobody need be ashamed of setting it on the table with the Château Haut Brion of the same year. Indeed, I am not sure that I would not name three Graves as the best wines of that vintage: Château Haut Brion, Domaine de Chevalier (like the Malartic, a Léognan) and Malartic Lagravière. I think they ought to go in that order. They are all light wines, and therefore I am not sure how I should compare them with the Latour or Ausone of the same year; both very fine specimens, and certainly fuller than the Malartic, I think fuller than the Chevalier, and therefore really only comparable with the Haut Brion. But for delicacy it would be a close thing amongst them all. It would be like the Caucus race; everybody would have to have a prize. I would very much like to come across the Malartic 1929; it should be remarkably fine drinking.

Château Gazin (not to be confused with the great Château Gazin in Pomerol or with Château du Gazin, a Fronsac wine that I have never encountered) is another red Graves that might be better known. In my mind I link it with Château Dauzac, a wine that has never disappointed me [CAMPBELL: *Yes, the Dauzac, property of the Johnston family, never presumes, but is sound and good and deserving of a higher place in the classification.*]; my experience of Gazin is much more limited, but that limited experience is entirely favourable. I wonder whether there is any connection between the vineyard and the General, who commanded one of the unconquerable divisions in Davout's Corps?

I think that I have now mentioned all the vineyards in the Graves district of which I have had personal experience. But it is no harm to observe that I have always found 'Graves de

Portets' a label by no means to be despised. It means no more than 'Pauillac,' or 'Margaux,' or 'Saint Julien'; but for some reason the district wines, sold without the name of a particular vineyard, seem to have unusual excellence in this locality. A few years ago I had a 'Graves de Portets 'of 1929 as a *vin ordinaire;* I never produced it without receiving a compliment upon it. It was introduced to me by one of the most charming men I ever met, even in the wine-trade; the late Henry Brooke, of J. H. & J. Brooke and Sons. [CAMPBELL: *I re-echo Maurice's eulogy of Henry Brooke, the saunterer, who was at Dulwich with me and was popular with masters and fellow-students, a saunterer but a good footballer.*] He had all the outfit of a wine-merchant; in the first place, you could discern the gentleman at a mile. He had a delicate sense of sauntering, a most engaging quality in itself, and almost an essential part of the make-up of a wine-merchant. To him, wine was the subject of one of the Arts; what his commercial assessment was, I cannot tell, but 'His soul was of the saints, And Wine to him was joy,' if I may so transpose my Lionel Johnson, whom he so much resembled. To see his nose delicately twitching over a glass that was the subject of a test was an education in itself. I was not surprised to learn that he and Frank MacKinnon used to spend their holidays together, walking and talking; for both had the love and reverence for beauty and tradition, although I am afraid the wine-merchant never persuaded the Lord Justice beyond Brown Sherry. He died unexpectedly a few years ago; and Paradise was enriched. Happily, his son, Cecil, verifies the title of the Company.

We now turn to Saint Emilion. I can never understand the lack of enthusiasm in Bordeaux for the wines of this district; I hesitate to think that it comes from selfishness, although most of the great houses have a proprietary interest in one or more vineyards of the Médoc, and few own estates in Saint Emilion. But few have estates in Graves; and I have never heard Graves decried in Bordeaux. When Saint Emilion is mentioned on the Quai des Chartrons it is generally in a semi-apologetic tone, with a *nuance* of disapproval. Yet the fact is that the wines of Saint Emilion are of a much more reliable character than those of the left bank. They may rarely attain the peaks reached by the great wines of the Médoc; but even a bad vintage gives a wine of marketable quality. The *terrain* is completely different. On the opposite side of the river, one finds a plain, with nothing more than undulations to break its monotony. Saint Emilion is broken country, with hills and dales, well timbered and watered, and full of variety. It is the romantic district also; here it was that the last of the Girondins hid in subterranean caves from their pursuers; it was hence that poor Barbaroux set forth on the vain pilgrimage that was to end in death. Although this is a book on wine, it is also meant to further conversation. [CAMPBELL: *He is quite right. Discussion and criticism of wine leads to conversation on all sorts of subjects and Maurice was a good bell-wether.*] Let us therefore take a glance at History.

In May, 1793, it was clear that the influence of the Girondins was eclipsed; their proscription followed. They escaped from Paris, hoping to find in Calvados or in Brittany a rally-point; at first they appeared to have found it at Caen, and if a citizeness of that town had not taken it upon herself to anticipate the methods of Chicago by 'bumping off' Marat, the history of France and the world might have been different. The effect of Charlotte Corday's act was to frighten the Caennais; the presence of the Girondin leaders was compromising; they persuaded these latter that Brittany would be a more favourable recruiting-ground; and the twenty-two young men enrolled themselves in the battalion of Finistère which was about to start for its depot.

The journey was not without excitement; news of their plan had reached the Government party, and at various towns upon the route steps were taken to intercept them. Individual members of the party broke off and went their own ways to their destruction; but six of them succeeded in gaining Quimper, whence they took passage by sea to Bordeaux. Alas! they found the southern capital dominated by the Terrorists, daunted and apathetic, quite unable to help them. But Guadet was a native of Saint Emilion. There, he felt sure, an asylum was awaiting them. He had a sister-in-law there to whom he determined to address himself. Nor was Madame Bouquey unworthy of his confidence. They have named the principal north-and-south street of Saint Emilion after poor Guadet; on the left you will find a tablet on the wall of Madame Bouquey's house, recording the fact that in the well beneath the building she sheltered seven of the proscripts. You may stand in the garden and drop a stone down the well; you will hear it echoing at a great depth before it finally splashes into the water. Look closely at the stone sides; you will see a series of rough and slimy holes that would be flattered by the name of steps, and yet that was their designated purpose. At a depth of some twenty feet, too overshadowed to be easily discernible from the surface, there is an archway, opening into a sort of cellar, below which is another room of the same size. There was an alternative entry from the house; to-day a stairway makes easy the visit of the tourist, but in Revolutionary days this entry was heavily concealed by furniture, and it was by the well that the refugees made their approach to their asylum. Being of some corpulence myself, I cannot but breathe an extra sigh for poor Barbaroux, who also possessed *embonpoint*. And there for five months this heroic woman succoured seven proscripts. To-day a short visit to the cellar even in the height of summer chills the marrow of one's bones; it is probable that the cave was only used when the sans-culottes were engaged in domiciliary visits. Nevertheless, the captivity cannot have been agreeable; and it is not surprising that by November the prisoners were so sick of it that they determined to escape to other climes. They separated; Guadet counted on the hospitality of a former client whose honour and fortune he had saved; but when, exhausted and in high fever, they knocked hopefully upon this door, it was shut peremptorily in their faces. Guadet and Salle returned to Saint Emilion, where Guadet's father received them and helped them to contrive a hiding-place; Louvet determined to walk on towards Paris, and, miracle of miracles, he was the only one who did succeed in saving his life. Pétion, Buzot and Barbaroux returned to Saint Emilion after vague and uncertain wanderings; with the assurance of those who have once received help from a particular quarter, they assumed that Madame Bouquey would succour them once again. But Monsieur Bouquey had returned from Paris; he forbade his wife to run any more risks; and the most that she was able to do for them was to persuade the local barber to receive them as guests. There is no better example of selfish vanity than the conduct of Salle, who kept sending his poetry to Buzot and Barbaroux for criticism, and Buzot and Barbaroux, who kept replying in lighthearted *jeux-d'esprit*, making it certain that their messengers must sooner or later expose their *cache*.

On June 16th, 1794, the house of Guadet *père* was surrounded, and before long the hiding-place of the younger Guadet and Salle was discovered. They were arrested; but alas! the papers found incriminated the Bouqueys, to say nothing of the elder Guadet. All these perished by the guillotine in Bordeaux. A grimly humorous touch was supplied by Salle; for the machine refused to function when the executioner tested it, and its repair was beyond his skill. 'Let me see,' said Salle, and promptly discovered and rectified the fault, so efficiently that his head fell in the basket a minute later.

Meantime, savage dogs had been turned loose in the subterranean passages of Saint Emilion, where it was believed that the other three proscripts were hiding. These watched the horror of the chase from the barber's window; but as soon as the dogs had been called off, they set out on foot, as any other place must be at least as safe if not safer. But Barbaroux had the bad luck to run into a party of troops; they happened to sound a call as he approached, and he assumed he had been recognized. He fired his pistol into his right ear, but failed to kill himself. A week later he was guillotined. Pétion and Buzot had better aim. Their bodies were discovered, blackened and in corruption, in a copse where they had said farewell to life. Pétion's was a dramatic story. The darling of the mob, it was he who had been chosen to fetch back the Royal Family from Varennes; and as he rode in the coach by the side of Madame Elizabeth he was vain enough to think that this saintly lady had fallen in love with him, and was making indecent advances to him, which, to her horror, he sought to return. After a year of miserable exile he died by his own hand in a Girondin meadow, with none but the ravens to attend his funeral. Nor has History given him that tear which his sufferings deserved, and which is so readily shed for the royal lady whom he so misunderstood and insulted.

So visit Saint Emilion with an ear for the tread of its tragic ghosts. Visit also the monolithic church, hewn out of the rock by the saint's disciples. It is under the modern church, itself worthy of a visit; to call the upper building 'modern' is almost an abuse of the comparative, for it dates back to the twelfth century, but the lower shrine is four hundred years older. Near Saint Emilion's cell is a crystalline spring which fills a deep basin; if you look down into it you will see that it is full of pins. The young women of the district throw two pins into the water on the Saint's day; if the pins fall athwart one another the lucky thrower will be married within the year. So that the tourist has plenty of interest in Saint Emilion besides the wines he finds there.

And he finds the wines even in the heart of the town; opening on to the square in front of the parish church are the gates of Clos Fourtet, itself one of the finest growths of the district. I first met Clos Fourtet in a vintage for which I have always had a great weakness, although it was supposed to be a mildew year—1912. I could name some twenty or thirty vineyards whose 1912 wines have given me pleasure; but most of them were light in texture and lacked body. The Clos Fourtet was a giant; it compared with the Latour. Both were admirable wines, and the only unsatisfactory bottle of the Clos Fourtet that I can recall was one which I decanted and took with me when crossing to Paris one Easter Saturday: 1926 I think it was. I lunched off sandwiches; I had been a teetotaller for Lent, so that the fault was as likely to have been with me as with the wine. I next met it in the 1917 vintage: one that I never much cared for, but the Clos Fourtet was very good. The 1924 I can praise without reserve; a fine wine of a fine year. And I found the 1929 very fleshy and good, too.

I have never quite agreed with the primacy generally accorded to Ausone amongst the wines of Saint Emilion; it was the favourite wine of King Edward VII and therefore needs no bouquets from me. But I am grateful to it for many a good bottle. Without going to look it up (and I am much too comfortable at the moment to do that), I cannot remember ever having drunk Ausone older than 1899; but in that year it was a model of silky finesse, and could keep on revealing new charms as long as the glass was refilled. I have always heard it said that the 1900 was still better; but I never had the good fortune to meet it. In the 1905, 1909, 1912, 1914, 1920, 1923 and 1924 vintages I have often saluted it; the 1909 was a very splendid wine, and the 1923 was an elegant amongst the elegants. Indeed, there is no need to employ the past tense here, for I have drunk it recently, and it is still in the prime of life.

I confess to some difficulty in writing about the Saint Emilion wines, for no proprietor is content to call his wine anything but a first growth (there are seconds, but nobody would ever come across them in England); and there is also a repetitive tendency in the names which makes it difficult to remember with guaranteed accuracy one's adventures in the district. The names Canon and La Gaffelière repeat themselves in all sorts of combinations, and there is a Château Canon La Gaffelière amongst the rest which calls for considerable respect. But of this group I have found the Château Canon *tout court* the most reliable. I never visited it, but I believe it is situated on a plateau near the town where it adjoins several of the most famous Saint Emilionais. It made a noble wine in 1906, which, with the exceptions of Ausone (of which I have no special memory) and Cheval Blanc (excellent, as always), I think is the only specimen of a Saint Emilion of this year that I have ever met. And it certainly was a formidable competitor to contend with the great 1906's of the Médoc and Graves. And the wine seemed to be free from the flavour of burnt grape-skin which has disfigured some of the other famous wines of that hot year.

The Curé Bon La Madeleine is another of these famous vineyards, with a pretty reason for its name. A Curé of the region once spent himself on his parishioners so extravagantly that it became idle to hope that his stipend could support him. But God blessed the vines in his vineyard, and made the wine so good, so reliable and so plentiful that he was able to make ends meet by the profit that resulted. As the average output would appear to be a mere five tonneaux of wine, the Curé's needs must have been small. I think I only once met the wine, which must obviously be difficult to find; but on that occasion I did greatly admire it.

Château Pavie has had a strange history, and since I visited it eleven years ago many changes may have taken place. It was then owned by a sad young man who lived the life of a recluse with a very handsome dog, which he always described as his best friend; and he had succeeded in falling out with what for convenience I shall call Bordeaux opinion, and was determined to market his wine himself direct. Now, I am as much opposed to unnecessary middlemen as anybody can be; but Bordeaux has never struck me as a good field on which to wage this particular battle. The great houses in Bordeaux command the world's trade; no individual in London is going to order by post from the Château X or Y, no matter how world-famous it may be. Therefore, the owner of a vineyard must seek his customer through the Bordeaux merchants. Why should he not deal with them direct? Just think what that would mean. Each great house in Bordeaux would have to employ a number of buyers, who would have to visit every vineyard month by month to see how the crop was faring; for they must report on the cheap little wines from the *palus* as well as on those of the great estates, as the shippers have to make up their standard cheap wines, which must necessarily be a blend of very small producers' output. What in fact the merchants do is this: they recognize that they must delegate a part of their business to specialists, and they go to the *courtiers* whose business it is to keep a perpetual eye on all vineyards no matter how humble. If Eschenauers find that they cannot repeat this year a purchase of 5,000 gallons of a little Fronsac wine, because the mildew or some other disease has ruined the vineyard, they just ring up their particular *courtier:* 'You remember that Fronsac we got last year? There is none available this year. What can you suggest as the nearest match?' And the *courtier* looks up his books and finds that he has something from Bourg that will make a good substitute; and so the thing is arranged. Now, the *courtier* naturally has to be paid; but if he were not there the merchant would have had to enlarge his own staff, and I do not see how this particular middleman could profitably be cut out. So that I suspect that the young proprietor of Château

Pavie had manufactured an unreal grievance; and I hope he has long since discovered his error and had the courage to acknowledge it. [CAMPBELL: *It all looks so easy, but Maurice was obviously non-conversant with Bordeaux business methods.*]

I have not often met Pavie, although its 1912 was amongst my loves, and it can still (or could quite recently) be drunk at the Queen's Hotel, Birmingham, quite sweet and full of merit. I should like to taste a sample of a more orthodox year. By the way, this is one of the names that confuse; there is a well-known Domaine de Pavie-Macquin which is often found in England, and a less-well-known Château Pavie-Decesse, which also produces an excellent wine. Pavie would appear to be the name of the plateau upon which all these estates are situate; although my recollection of the Château Pavie does not include anything like a plateau, the estate being on pleasantly broken ground, with a romantic spring wandering through the garden. Another slope that gives its name confusingly to a number of leading Saint Emilions is that of Cadet; we find Cadet-Bon, Cadet-Piola and Moulin-du-Cadet as well as the Châteaux Haut-Cadet and Pavillon-Cadet. I do not remember meeting any of these except the Cadet-Bon, which did not register as outstanding.

But I have neither time nor space to chronicle all the good First Growth Saint Emilions; they run to dozens and keep a fairly constant high level without ever singing the inspired song. But let me spare a line to praise a little wine that has the courage to offer itself as a mere Second Growth: a distinction in this area! I refer to Château Haut Simard, an excellent little wine that will bring discredit to no table and can usually be drunk without any sense of extravagance, as the price which it commands is little above that of an *ordinaire*. But I have frequently quaffed it with pleasure; and it deserves as much praise for its excellence as for its modesty.

We now come to a district that is usually classified separately from Saint Emilion proper; it has the gravelly subsoil that helped to make the fame of the red Graves, and it is known as Graves-Saint-Emilion. Its wines borrow the excellences of the adjoining districts of Saint Emilion on the one side and Pomerol on the other; they have attained a delicacy comparable to that of the Médoc wines, while at the same time maintaining the *hardiesse* of the right bank. Amongst these vineyards that of Château Cheval Blanc is *facile princeps*. In my own time I have seen the reputation of this great wine advance by leaps and bounds; always esteemed, it now is reverenced, and if it continues its upward career it will ere long be recognized as the greatest of all Clarets. I have never come across a pre-phylloxera Cheval Blanc; the oldest I have met was, I think, the 1899, a very beautiful wine. But the first time I recognized the greatness of the vineyard was when I accidentally encountered the 1914 at *Les Gourmets* in Lisle Street; and here, I thought, was the authentic note of quality. Stephen Gwynn shared my admiration for this beautiful wine; indeed, I rather think that it was to him I owed my introduction: one of many good turns. Since then I have met most of the better vintages down to 1928, as well as going back to the 1909 and 1906; and the wine has never failed me.

But just as Mr. Baldwin achieved fame in a night by the address he delivered in the Middle Temple Hall entitled, 'On England,' so did Cheval Blanc conquer the world when it set before us its wine of 1921. I have often said that the most *remarkable* Claret I ever met was the Château Haut Brion 1906, which poor Hew Brown once described as 'a first-class Claret with a dollop of first-class Port in it.' In its prime (and there are good bottles surviving to-day) it was a mighty wine; it had more of the virility of a Burgundy than of the delicacy of a Claret, and it was anything but a typical Bordeaux. But it was great. It had majesty and dignity; but it was the majesty and dignity of Henri Quatre, and with the right people it was prepared to be 'Hail fellow! Well met!' It will be remembered that 1906 was a very hot summer; and in the Haut Brion, as in many other wines of that year, one could detect the flavour of the scorched grapeskins, a flavour that will be familiar to those who know their 1921's and 1926's. Now, in 1921 Château Cheval Blanc made a wine that overtopped the Haut Brion 1906 [CAMPBELL: *It is still possible to put the* 1906 *Cheval Blanc up against the* 1906 *Haut Brion and this would be a fairer and more interesting competition. Both are still very good, but the Haut Brion seems to be lasting rather the better. The* 1921 *Cheval Blanc was almost a freak and* hors concours. *If a bottle of it comes your way my advice is to grab it with both hands*]; because, made in as hot a year, and equally virile and full-bodied, it was free from the burnt-skin flavour, and was as elegant as it was powerful. I once described the intrusion of the Haut Brion amongst a lot of classic Médocs as Napoleon walking in amongst the Bourbons. The Cheval Blanc calls for a more refined exemplar; perhaps when Gustavus Adolphus landed in Germany he was supplying the image. I think I must dethrone the Haut Brion and replace it by the Cheval Blanc; for this is certainly in every way the most remarkable Claret I have ever drunk.

And the joy about Cheval Blanc is that we have not finished yet. Each new vintage produces new excellences; even in years that have not succeeded elsewhere this remarkable vineyard pursues its untroubled way; and even its 1925 was a wine worthy of attention. I wonder how it fared in 1927 and 1930 and '31, when so many wines failed. The Château itself is a charming residence where the present proprietors are in constant attendance; they have no interest in life save the improvement of their wonderful estate, and they give personal attention to every vine that is contributing to their fame. In such circumstances we may hope that the upward curve has not yet reached its peak, and that for many years to come Cheval Blanc will continue to be one of the most splendid jewels in the brilliant crown of Bordeaux.

Next door is Château Figeac, somewhat overshadowed by the glory of its neighbour; but there have been vintages in which Figeac has been no unworthy competitor, and higher praise could not be given. Figeac is another name well scattered through the district; there is a Latour-Figeac, a Latour-du-Pin-Figeac, a La Tour-du-Pin-Ströhl-Figeac, a Domaine de Figeac, a Grand-Barrail-Lamarzelle-Figeac, both a Château and a Clos Ton Figeac (the Château wine has given me several very good bottles) and a Château Cormey-Figeac; yet I have not discovered whence the common name comes.

One portion of what is loosely called Saint Emilion is more correctly classified apart, as its wines have characteristics of their own and are worthy of separate study. I refer to Pomerol. The village of that name numbers less than 1,000 inhabitants, but the wine area is fairly extensive, and includes many famous vineyards. The Pomerol wines at their best are endowed with a delicious perfume, which some liken to that of violets and others, more greedily inclined, to that of the truffle. I can see what each party is driving at, although neither quite gets there. The violet when freshly gathered still has the earth about it, as has the truffle; is it the earthy bouquet, duly sweetened, that each side is recognizing? At any rate, the perfume of a Pomerol is powerful and pleasing and full of promise that is always performed. There's alliteration for you, as Humpty Dumpty might have said!

The wine itself is usually dark and full, and is rather sweeter in flavour than other Clarets. It is a sappy wine, one to smack one's lips over, each sip revealing something new and agreeable. Anybody who had the good fortune to meet the Château Pétrus 1917 about ten years ago will know what I mean. Whether it has held I cannot say; they used to have magnums of it at the New University Club which perhaps are still to be drunk at the United; and in magnums the wine should certainly have held, although André Simon looks on Pomerols as wines to drink rather than to keep. Yet Charles Berry once gave me a beautiful Vieux Château Certan that was a pre-phylloxera (an 1877, if my memory serves me), and at something like sixty years of age it was a lovely and very gracious thing. Its welcome had the dignity and charm of that of a dear old Reverend Mother in a very exclusive convent. I have never before associated either truffles or violets with holy nuns, so perhaps I shall have to revise my analysis of the perfume!

Besides the two vineyards I have mentioned, one finds in Pomerol l'Evangile, La Conseillante (which I think I have never met), Trotanoy and Gazin, not to be confused with the pleasant little red Graves of which we spoke earlier. These six Châteaux are probably the best representatives of Pomerol, where, however, many more vineyards are to be found living up to their claim to be Premiers Crus. I once had the luck to possess some of the wine of l'Evangile 1924. I got it from Messieurs Avery of Bristol, and when I say it was one of my luckiest purchases from them I am putting it very high. It was a beautiful wine; in a strange way it suggested the lusciousness of the blackberry; aromatic, full and distinguished, it earned me many an agreeable compliment. I think that it and the Pétrus 1917 were the only Pomerols I ever cellared myself. Oddly enough, 1917 was a very unsatisfactory year in most districts; but the Pomerol wines were all successful, and, although I have doubts of their ability to stay the course, they had every other attribute that wine should have.

There is no other district in Bordeaux that would repay a special study; but there are two vineyards in humbler parishes which I venture to mention by name, for different reasons. The first is a wine of Fronsac, and only a Deuxième Côtes at that. And I have never yet succeeded in finding a bottle, although I should like to drink the wine of a gallant soldier, comrade of so many of my own comrades in the 29th Division. For Château Pontus is the property of Général d'Amade, who commanded the French Expeditionary Force at the Dardanelles; and I am sure that he would be the first to acknowledge that some at least of the courage, the tenacity and the kindly charm that endeared him to all who met him in the field came from the wine of his own vineyard. I wish we could have it bottled under the sign of the Red Triangle that was our own badge, and that at our reunions it should fill our glasses as we drank to the days that are gone and the friendship that has endured.

The other Château made one of the best jokes ever associated with a British railway.[1] I suppose there is no lover of wine who has not wondered at one time or another how any decent Claret can survive in the locker of a railway dining-car, bumped and jolted in every direction, and yet in some way managing to make a respectable appearance in the diner's glass. One day the caterer for the Great Western Railway found an estate that seemed by its name to have been created to fulfil his purpose. And for years afterwards the traveller who sought a glass of Claret to solace his transit to Plymouth or Bristol or Birmingham found at the head of the list the Château Mille Secousses! There really is such a Château; it is in Bourg, and its full title is Château Mille-Secousses-Chenu-Lafitte, and it is owned by the Count and

[1] See p. 249.

Countess Péan-Oberbech-Clausen, who also own Château Rider-Chenu-Lafitte next door and Château Gourdet-Chenu-Lafitte a little further on. The three vineyards produce about eight hundred tonneaux of red wine which is classed as 'exceptional,' and they sell it all under the label of Mille-Secousses. The part of the name that is common to all three vineyards commemorates the son of the Countess, René Chenu-Lafitte, who died in battle in the last war. One reverently thinks that there could be no more noble memorial of one who has died for mankind than the consecration to his memory of the juice of the grape.

Here, then, is my praise of Claret. When formerly I wrote upon it Monsieur Latheuillière, the editor of Cocks et Feret's *Bordeaux et ses Vins*, most kindly corrected several of my errors, and encouraged me to borrow from his great work, which is the 'Bible' of Bordeaux. A great deal of the detail in this section derives from this kind permission to make use of his book, to which I have only once turned in vain for information. And perhaps in the next edition, due very shortly, I shall be able to find out all about Château Petit Bal, which I believe to be a Saint Emilion, which produces a very agreeable wine, but which for some reason fails to make any appearance in this otherwise exhaustive guide to the wines of Bordeaux.

BURGUNDY

EVERY journey I make by Underground is the source of a certain amount of irritation. I find myself faced with elaborate quotations from George Meredith in praise of Burgundy; and these appear on the advertisements of a wine that has just as much right to call itself Irish Whiskey as to take the name of Burgundy. Closer examination reveals that the article advertised comes from Australia, or from some other part of the Empire; and I am prepared to believe that it has been made honestly, in the Burgundy manner, from Pinot grapes. It is a perfectly genuine wine, straightforward in everything but its name and its attempt to attribute to itself what Meredith said about Burgundy. For it is *not* Burgundy; and in fact it only faintly resembles Burgundy. Mind you, for those who like that kind of wine, it is just the kind of wine they would like. [CAMPBELL: *Many a true word is spoken in jest. The rather severe orthodoxy of mv own views on the nomenclature of wine is, I think, well known. Maurice can hardly have been ignorant of the fact that, as the result of a lawsuit settled in the Courts in this country, the name Burgundy can be used for Australian wines with certain provisos.*] It is strongly tonic, for it is very ferruginous; but only once, and then by a rare accident, did I succeed in finding an Empire wine of Burgundy type that had been allowed to age; and at ten years old it did taste rather like Burgundy, although not a very good Burgundy. When Meredith went into dithyrambs about Burgundy he was speaking of one of the noblest and greatest of red wines, that has gladdened the hearts of generations of mankind. That wine is made in France, and nowhere else; it comes from a very strictly delimited area in the neighbourhood of Dijon. When Burgundy was a Duchy, this area was at the heart of the Duke's pominions. His rule varied from generation to generation;

there were times when he also ruled over Flanders and the Low Countries. The inhabitants of these parts were his richest subjects; and they were able to pay for the very best of the wine made each year. They created a tradition that has lasted to our own time; and the seeker after the best Burgundy to-day[1] will be well advised to travel to Belgium or Holland. In England there are certain wine-merchants that have the Burgundy tradition, and one can count upon them to produce something good; but for one really good bottle of Burgundy sold in this country there are fifty mediocre and a hundred bad. I can only remember two lots of Burgundy in my own cellar that were outstanding. One was a magnificent Clos de Vougeot 1904 that I got from Christopher's about 1925, the other was a Les Bonnes Mares 1921 that came from Avery's of Bristol. The latter was a less expensive wine, because it was younger and the vintage year was not classic; but it really proclaimed the true merit of Burgundy as loyally and satisfactorily as its elder superior in rank, which, however, was a superb wine: probably the best I ever had in my cellar. For let there be no doubt about it: Burgundy at its best overtops Claret at *its* best. [CAMPBELL: *'Burgundy at its best overtops Claret at* its *best!' Again I must say 'de gustibus, etc.' Like the rest of us Maurice was susceptible to environment and his verdicts were often emotional and influenced by pleasant companionship. His 'best' Burgundies appear to have been worthy representatives, a Richebourg* 1923, *a La Tache* 1904, *and a Volnay, Caillerets* 1889, *and it would have been interesting to have known which three 'best' Clarets he would have put up to meet the challenge and acknowledge defeat.*] You will only drink four or five bottles of truly first-class Burgundy in your whole life (and you will be lucky if you find so many; only three have rung that bell with me; good as the 1904 and the 1921 were, none of them was the really outstanding thing I have in mind). But you can drink Claret of the highest class several times in the year: Claret that should be drunk kneeling, with every sip consecrated as a libation to Heaven. I wonder if I can get an image from the orchestra. Everyone knows how tender and beautiful the flute can be in a solo part; but whosoever has had the luck to hear the solo played

[1] No; not to-day. But to-morrow!

by the recorder recognizes a fullness, a roundness and a pathos of tone to which the flute cannot attain. Now Claret flutes for us all the time, while Burgundy is usually content to grate on its scrannel pipe of wretched straw. Occasionally it, too, will flute with grace and tenderness; but, once in a blue moon, it will produce the recorder, and then, indeed, our hearts are melted. Such a moment occurred to me the other day, when at the hospitable table of my good friend, Sir Francis Colchester-Wemyss, I partook of a bottle of Richebourg 1923 that fulfilled all the conditions that govern the best of Burgundy. I had honestly forgotten how good Burgundy could be, and this was like a new revelation.

Now, why should Burgundy be so uncertain and difficult when Claret is so easy and reliable? The answer is to be found in history. It is probable that the vines of Burgundy would never have burgeoned into so great a fame if it had not been for the fact that nearly all the great vineyards were founded by communities of monks, ensuring a continuous tradition of skilled attention to the vines and the vintage. In order that a wine may be relied on year after year, it is necessary that the whole vineyard should have the same care and that there should be continuity of practice. A vineyard belonging to Jean this year and to Jacques next year and to Paul the year after would not be likely to produce a wine of the same constancy as that produced under the continuous care of Robert. In the past, a community of monks always proved the most satisfactory nurserymen; there was no sudden change of ownership brought about by death or alienation, and the tradition of the vineyard passed from hand to hand almost imperceptibly. Under such care the great vineyards of Burgundy achieved their excellence; around them were grouped smaller estates, whose owners were not above learning from the monks, and who gradually spread the art of wine-making far and wide throughout the district. Now, to make good Burgundy, three things are necessary: a good vineyard, incessant toil and care and the Pinot grape. For hundreds of years the lover of the wine of Burgundy has been making war on a determined interloper, called the Gamay, which is a vine that produces a much more plentiful crop of grapes that make a highly perfumed wine. Where the

Pinot would give you a hundred dozen of aristocratic wine, the Gamay would probably give you more than one hundred and fifty dozen of the demi-mondaine variety. And, just as there are a large number of persons who cannot judge as to whether the lady to whom they are talking is an aristocrat or a demi-mondaine, so are there an even larger proportion of persons who cannot discriminate between a wine made from the Pinot and a wine made from the Gamay. So the artist continues to plant the Pinot; the money-grubber plants the Gamay. Now, as long as the great vineyards were in the hands of the religious orders, principle prevailed over interest, and a wine from those vineyards could be relied upon to be Pinot. But the orders were driven forth, first, in the French Revolution, and again, in the case of those who had recovered their property, at the beginning of the present century. In Bordeaux, whenever one of the famous estates came on the market, a syndicate, largely supported by the family who had owned it, bought it as an entire vineyard, and continued the tradition. In Burgundy, unfortunately, the great vineyards were broken up and sold to a series of little proprietors. The Clos de Vougeot is parcelled out amongst more than forty owners; the same story can be told of most of the other famous Clos. Now, every one of these little proprietors is entitled to sell his wine under the name of the world-famous estate; it would be too much to expect from human nature that amongst so many some should not be found prepared to sacrifice quality to quantity, and to substitute the Gamay for the Pinot. In the case of the Rhine wines, it is customary to proclaim on the label where possible that the wine has been made from the Riesling grape (which, by the way, is supposed to be the same grape as the Pinot of Burgundy). It would be an excellent reform if the Burgundians saw to it that a bottle which carried the word 'Pinot' on the label was guaranteed to be made from the unadulterated juice of that grape. I have never seen a bottle so labelled, but I see no obstacle to the reform.

And labels have been the cause of much trouble in France, and in particular in Burgundy. There is a French law (of course, not applicable to this country) by which the contents of a receptacle must correspond in origin with what is stated on the label. Australian Burgundy would probably have some difficulty in France, for Burgundy is a place-name; there is no place called Burgundy in any part of the world except France, and a label 'Burgundy' can only be put on a bottle containing wine coming from Burgundy. A very sound law, you will say; but wait until you see the way that foolish vanity has made it an instrument of deceit. Nuits is a little town in Burgundy; indeed, the Burgundy district is mainly divided into the Côte de Beaune and the Côte de Nuits. The wines of the Nuits slope are those with staying power; the Beaune wines are not so lasting. Almost all the most famous vineyards are on the Côtes de Nuits. In the commune of Nuits itself there is one vineyard which for generations has been famous for the excellence of its wines. It is called the Clos Saint Georges. So proud were the inhabitants of Nuits of the fame of their great vineyard that they petitioned to be allowed to tack its name on to the name of their parish, which thus became Nuits-Saint-Georges. But observe, before that change, Heaven help any little *vigneron* who sought to call his wine by the name of Saint-Georges; the highest title he could give it would have been 'Nuits.' But the moment this change had been made he could with perfect legality and propriety call his wine 'Nuits-Saint-Georges,' which was also the right of the owner of the Clos. I have found when sitting on wine-committees that venerable old gentlemen, strong in the best wine tradition, have always expected the label Nuits-Saint-Georges to cover a first-class Burgundy; and I suspect that in the old days the Clos used to sell its wine as Nuits-Saint-Georges. But to-day such a label does not say more than that the wine is a product of Nuits; and, as there are fifty or sixty vineyards in the commune, most of them of very inferior rank, the probability will be that a bottle of Nuits-Saint-Georges is an *ordinaire.*

The other communes followed the bad example of Nuits. Chambolle got leave to call itself Chambolle-Musigny, after its famous vineyard, Les Musigny. Gevrey tacked on the great name of Chambertin; Vosne gave itself the addition of Romanée. It cannot be too strongly emphasized that a Vosne-Romanée is only an *ordinaire,* and a remote cousin of the three great bearers of that title: Romanée, Romanée Conti and Romanée Saint Vivant. And the Clos de Conté is no relation of these either; nor can I supply information as to where it is situate. I can remember a hospitable friend of mine who promised me a good bottle of Burgundy. He told me that it was not merely a Chambertin, it was a Gevrey Chambertin. He was unaware that he was proclaiming his bottle an *ordinaire* and not a prince of the blood. The white wines have suffered in the same way; a Puligny-Montrachet is *not* a Montrachet, but a Puligny.

Now, with all this chicane on the subject of vines and labels, is it any wonder that the task of finding a genuine bottle of fine Burgundy is infinitely more difficult than the finding of a good bottle of Claret? And it does not stop there. The Burgundy shipper who wants to sell on the English market must probably deal with a wholesaler, who will have several retailers to supply, so that he must not show favouritism between one and another. Now, the average retailer who wants to stock a particular wine will probably want to take at least twenty dozen, so that the wholesaler will probably want about four or five hundred dozen. Well, there are very few proprietors in Burgundy who are prepared to supply so much from a single vineyard. I remember with the greatest pleasure a Burgundy given to me by André Simon some years ago. It was a 1926; it came from a little vineyard in Pommard called Les Epenottes. But the total output of that vineyard was not more than about seventy dozen bottles a year. Such a quantity would be useless to the average wholesaler for the reasons just given; so the proprietor would be driven to one of two courses: either to find a person like André Simon (and I can imagine no more difficult task!) who would take his *totalité* and break it up amongst his friends by sale or gift, or else to sell his output to a merchant of Dijon or Beaune, who would blend it with similar parcels of the same year and market it under the label 'Beaune Supérieur' or 'Pommard Supérieur' (for it is sometimes forgotten that the labels 'Beaune' and 'Pommard' are not mutually exclusive).

Have I now made clear how difficult it is to get a star Burgundy that shall be as straightforward and unmixed as a Château Lafite or Château Latour? Into this sea of confusion all the rogues have plunged, and they batten on the uncertainty so created. I suppose that if I could get to know the owner of a parcel of the Clos de Vougeot that I could be sure was an honest producer of Pinot wine, and learn from him the channels through which his wine would reach the customer, I could be sure of getting pure, first-class Burgundy. In the same way, if I got to know a reliable Burgundy shipper in Dijon or Beaune, I suppose I could rely on the truth of the label on any bottle shipped by his house. But this calls for more time and erudition than the average citizen can afford. Who am I to appraise the honesty of a man living several hundred miles away from me, whom I have never met? But here on my own doorstep I have a choice of a number of wine-merchants whose reputation and skill are known to me, and whose business it is to deal with these problems for me; and I am quite satisfied that on the subject of Burgundy I should be a fool to back my opinion against theirs. Mind you, I hold to what I say; our London merchants (and still more does this apply to those in the provinces) do not seem to hit the bull's-eye as often in Burgundy as they do in Bordeaux. In particular, they seem to give a great deal too much rope to certain shippers, whose name I always associate with cooked Burgundies; and having said that I refrain from making any mention of individual Burgundy shippers, because I have had unfortunate experiences with some that are usually mentioned as absolutely reliable. But what can you expect, when Monsieur X's Chambertin is made from grapes grown in the first, tenth and seventeenth parcels of the vineyard, while Monsieur Y's comes from the second, fourth and twenty-second? This necessarily opens the door to blending; and once blending is permitted it would be trying shippers very high to expect them to refrain from imitating their colleagues of Champagne, Xeres and Oporto by topping up a parcel with 'something good' left on their hands. I am never happy about my bottle of Burgundy until I am sipping it. And in the circumstances I consider myself lucky to have come across so many good bottles.

The red wines of Burgundy come, so far as excellence is implied, from the south-eastern and eastern slopes of a low line of hills running south from Dijon to Chagny, known as the Côte d'Or. The heights are inconsiderable, as hills go;

they vary from about six to fifteen hundred feet above sea level, but much less than this above the adjoining plains. The slopes are divided into three principal stretches: the Côte de Dijon, the Côte de Nuits and the Côte de Beaune, of which the latter two are the most important. Along their eastern side runs the railway line from Paris to Marseilles; and passengers to the south who happen to look out of the right-hand windows of the train will see proud notice-boards identifying some of the most famous vineyards. I am sorry to admit that this is the nearest to the Burgundy vineyards that I have yet reached; the years of the financial phylloxera interrupted my studies of the wine-lands, which are not likely to be resumed until more important matters have been satisfactorily arranged and Peace once more settles on Europe. But my masters have been kind in their instruction, illustrated in the most appetising manner by samples of the Burgundian output; and I think I know enough to write a primer on the wine, if I cannot claim the same acquaintance with it as I have with the glories of Bordeaux. [CAMPBELL: *It is a pity Maurice wrote his chapter on Burgundy without having visited the Côte d'Or. Too much of his information is obviously hearsay and gives a misleading and biased impression: it is true he attempts to qualify his highly coloured opinions by asking in the previous paragraph, 'Who am I to appraise the honesty of a man living several hundred miles away from me, whom I have never met?' Yes, indeed, and like so many of his delightful countrymen he did not seem to mind whom he hit as long as he hit somebody.*]

Although the Côte de Dijon produces a large quantity of wine that would rank above mere *ordinaires*, it is not until we get down to the commune of Fixin, which borders on the Côte de Nuits, that we find a vineyard worthy of special attention. This is the Clos de la Perrière, remarkable for having preserved its unity through all the periods in which the vineyards of Burgundy were broken up amongst small proprietors. Originally formed by the monks of the Abbey of Citeaux, it has a pedigree of single ownership down to its acquisition by Monsieur Pierre Joliet, whose property it was twenty years ago when Monsieur Camille Rodier published his valuable book, *Le Vin de Bourgogne*, from which I have acquired a great deal of my information. The effect of unity of ownership has been very marked in the wines of this estate, which for years commanded the same price as Chambertin, and would tempt me to a higher bid than I would make for most bottles labelled Chambertin to-day. La Perrière is a lasting wine, of great breed; I have not often met it in England, but it is worth noting the name, unfamiliar to the average diner-out, in case you should stumble upon it by accident. The output is less than five thousand gallons a year, so that it is not a likely inhabitant of a London cellar. The neighbouring vineyard of Le Chapitre, also enjoying single ownership, is supposed to be a close runner-up to La Perrière, but I have never tasted the wine myself.

Gevrey is the first commune of the Côte de Nuits, and Chambertin is the first vineyard of Gevrey. That Chambertin enjoyed pride of place amongst red Burgundies one hundred and fifty years ago is, I think, suggested by the fact that it was the habitual wine of Napoleon, who carried a supply of it with him on all his campaigns, but (blasphemy!) drank it well diluted with water. On the eve of Jéna his wine-waggon had not arrived from Gera, so he drank the local wine, of which I have never heard in any other connection. But Napoleon was not a *gourmet*, and Chambertin has received more discriminating praise. It must be one of the oldest vineyards in Europe, for its history is traceable back to the seventh century. Originally it was cultivated to about half its modern extent, by the monks of the Abbey of Bèze, who gave it the name of the Clos de Bèze. Then an enterprising peasant named Bertin, who owned the adjoining field, thought that he might be able to turn this into a vineyard; and the produce of the Champ de Bertin soon overtook the fame of that of the Clos de Bèze. To-day the two properties have been united, but alas! have been parcelled out amongst twenty or thirty proprietors. The produce of the united vineyards, which cover about sixty-seven acres, cannot be more than about 25,000 gallons a year, or say about 150,000 bottles. Probably a hundred bottles labelled Chambertin are uncorked each night in London alone, where on this basis about three and a half times the total output of the vineyard is consumed each year. This method of performing the miracle of the loaves and fishes does not appeal to me.

On the other hand, it must be made clear that the vineyards in immediate proximity to the great estate have taken the liberty of adding its name. Latricières, Charmes, Mazoyères, Combottes, Mazis and a number of other names are indicative of an origin in neighbouring vineyards that have been classified as Deuxièmes Cuvées. They may bear the same relation to the great Chambertin as a Château La Tour Carnet would bear to a Château Lafite of the same year (and one knows that sometimes the lesser vineyard will overtop the greater); but they ought not to sell themselves as though they were really the produce of a part of the principal estate. There are also Troisièmes Cuvées, but I have never seen their names on an English wine-list.

Professor Saintsbury did not care for Chambertin, which he thought leaned towards coarseness. I have always found it a manly wine, but none the coarser for that. There is a robustness about it that one does not associate with wines from the more southerly slopes; it is, perhaps, the Ulsterman of Burgundy, and that is no mean compliment. One hears the clang of armour in its depths; Mozart closes his clavecin when it is poured, and Doctor P. G. Wodehouse defers telling that story he was about to begin. But Bach moves towards the organ; Henry Ainley opens the pages of *Henry V;* the host turns on the lighting of his Rembrandt; and the noble wine blends with them all. Aye, 'noble' is the word; and I do not know any produce of the grape that so well deserves the adjective.

Moving still southward, the next commune is Morey; and in Morey are to be found two entire vineyards and the part of another which have made the delicacy of Burgundy their special aim. The Clos de Tart and Les Lambrays are neighbouring estates, all of which are within this commune; a small portion of the vineyard of Les Bonnes Mares overflows in here from Chambolle next door. All three produce wines of unusual breed and distinction. The first two were founded by communities of nuns, not monks; and although Les Lambrays was divided into fragments at the time of the Revolution, both vineyards have had the advantage of single ownership, the one for most and the other for all of its history. I have only once met Les Lambrays, and rarely the Clos de

Tart; a 1904 bottle of the latter was amongst the most delicate wines I ever tasted. Les Bonnes Mares will always occupy a high place in my esteem because of the 1921 of which I have already spoken; but I have met it in other vintages with great, if not equal, pleasure. It has a perfume that suggests really good sealing-wax; I do not know if others will read this as the praise which I intend. The output of the Clos de Tart would be about 2,500 gallons; of Les Lambrays about 3,500 gallons, and of Les Bonnes Mares about 5,600 gallons. Dividing the gallons by three we get a rough estimate of the number of dozens produced; so that there are hardly 4,000 dozen amongst the three of them, to serve the whole world!

The great wines now commence to crowd, for Chambolle, Gilly and Vosne are communes that are very close together. We meet them in that order as we proceed southwards; and the continuation of Les Bonnes Mares is the first estate we reach. Les Musigny and Les Petits-Musigny are the only other first growths, and they are at the other end of the parish, almost neighbours of the Clos de Vougeot. I remember a lot of Les Musigny 1923 with which Messieurs Christopher stocked the 1st Lancashire Fusiliers' Mess about ten years ago, at a remarkably low price. It was all that good Burgundy should be; and it lost nothing by being drunk in such excellent and hospitable company. The day after I had first tasted it I made haste to seek a little for my own cellar; but others had been wiser virgins than I, and it was all gone. I was not surprised.

Amongst the second growths of Chambolle are Les Charmes and Les Amoureuses, which I have seen on some wine lists. I commend their honesty of label, for I cannot repeat too often that a Chambolle-Musigny bears the same relation to Les Musigny as a Margaux bears to Château Margaux. The total area of the two vineyards morally entitled to the great name is only about twenty-five acres, and the genuine wine is scarce.

Gilly was originally the name of the next commune, but it is more frequently called Vougeot nowadays, although Gilly has improved its own name to Gilly-les-Vougeot. The great Clos stretches right across the parish from Chambolle to Flagey; its 125 acres form the largest vineyard in Burgundy, and would that it had been maintained as a single entity. But it has been broken up amongst some forty proprietors, each of whom treats his own little bit of the vineyard as he pleases, knowing that he is entitled to sell his wine under the name of the great estate. I am told that the standard of honour amongst the proprietors has been high, and that none of them have planted the Gamay. Upon this point I have no personal knowledge, but I hope it is true. If it is, there have been some very fraudulent uses made of the label. How often have I not grimaced over a glass poured from a bottle the label of which assured me that it came from the great Clos! And I remember on one occasion I was served with a bottle of venerable (and I believe genuine) Burgundy, which I had ordered as Clos de Vougeot 1898; it was so good that I asked to see the bottle, in order to give the shipper a good mark; and the waiter brought, not only the bottle, which was so labelled, but also the cork, which was branded with the shipper's name and the word 'Chambertin'!

Clos de Vougeot is the lady amongst Burgundies; but she is a very majestic lady. I have never understood Professor Saintsbury's suggestion that the wine tends towards Bordeaux; it is much too full for that. Its perfume recalls that of cedarwood; it has a latent sweetness that is never obtrusive, and there is a supple firmness about it that suggests Diana. It is one of the four great Burgundies that constantly compete for the leadership. It would probably regard Chambertin with the air that Marie Antoinette might have adopted towards Mirabeau; it would acknowledge the ancient blood of Romanée Conti, without quite according the salute due to Royalty; to Richebourg it would behave as Mrs. Thrale towards Doctor Johnson. Of its own priority it would have no doubt. In the charm of the Presence I also have had no doubt; but at an impartial distance I am not so confident. It is odd that my three greatest bottles, being all Burgundies, should not have included a Clos de Vougeot. On the other hand, I have never owned a more magnificent parcel of wine than my Clos de Vougeot 1904; and, apart from the casual bottle drunk in a cheap restaurant, I have found the label prophetic of satisfaction.

I hope that the coming of another war has not interrupted the traditional compliment paid by the French Army to the great Clos. In the wars of the Revolution, as Colonel Bisson led his regiment past the vineyard, he called a halt, faced his men towards the vines, and ordered them to present arms and sound the general salute, 'for,' said he, 'that vineyard contains the blood of France.' The gesture pleased the nation, and since then every body of troops passing that way pays a similar compliment to the Clos. I once suggested that the units of the army in Eire might salute Guinness's Brewery in the same way; but the idea was not approved, doubtless for reasons of traffic, the Kingsbridge terminus being in the vicinity. It has been suggested that the Navy might care to pay a similar tribute to Bournville; but for the moment they are otherwise engaged.

I have never come across the white Clos de Vougeot, about five acres of the vineyard being planted with the Pinot Blanc and giving a white wine of considerable reputation. But it does not seem to reach this country, as its name is never to be found in our wine-lists. It is probably the opposite number of the white Haut Brion; if it is as good I hope to find some before I die.

The next commune of Flagey has only one great vineyard, Les Grands Echezeaux, to the twenty-three acres of which estate are attached the eight acres of Les Echezeaux de Dessus. Both of these estates rank as first growths, but I have never come across the lesser label, and I am unable to state whether both wines are given the accolade of 'Tête de Cuvée.' I have cellared Les Grands Echezeaux with great satisfaction, although I would not rank the wine amongst the very greatest of Burgundies. Indeed, I was once asked whether I would object to receiving it labelled 'Beaune Supérieure'! The merchant in question had bought more of the classed wine than he could readily get rid of, so he used a considerable amount of it to improve the reputation of his *ordinaires* and only increased the price of that portion which he sold under its proper label. I tasted the contents of both bottles, and the modest label did not chafe the proud spirit it misproclaimed. The wine was again what I call 'sealing-waxy,' and had a good bite to it; it was an admirable accompaniment to a partridge.

We now reach Vosne, or, as it is now called, Vosne-Romanée; and surely never did merit crowd itself into so narrow a space. With your back to the railway, as you leave the village, you have in front of you the twenty-three acres of Romanée-Saint-Vivant; behind that vineyard on the left is Romanée-Conti, amounting to about five acres, and behind that again the single acre of Romanée; while on the right of these, also behind Romanée-Saint-Vivant, is Richebourg, extending over about twelve and a half acres. To the left of Romanée-Saint-Vivant, beyond the less important first growth of Les Gaudichots, which extends to about fourteen acres, is the four-acre vineyard of La Tache; and, as so often in Burgundy, it may be said here that the best goods are to be found in small parcels. In September, 1926, I spent a few days in Paris; and I decided that I would try the merits of Lapré, in the Rue Drouot. I went early in the day to leave my order, and I ingratiated myself with the *sommelier* by ordering my wine first and then choosing the food that would show it at its best. The first wine was a Meursault; I cannot remember the vintage, but it was followed by a La Tache 1904—oh! what shall I say about it that can give any idea of its excellence? So marvellous, so delicate was the bouquet, it seemed an impertinence to go further and taste this miraculous liquid; and yet, on the palate it almost made one regret as a waste of time the moments spent in taking in the perfume. This indeed was one of the Three Bottles; its excellence was beyond all description or recapture; yet I have never seen a good vintage year of La Tache on a list since without making the vain effort to call back the past.

But La Tache must take a place junior to the Romanées and Richebourg. Of the three Romanées, the Romanée Conti is usually accorded pride of leadership; but I have drunk bottles of the others that make me wonder how much better any wine in this world could be. The venerably classic years have not come my way with any of them; nothing older than 1911. I have had my share of 1911's, 1915's, 1919's and 1923's, and there are no sad memories. When one thinks of Bertrand du Guesclin the image is as much of good manners and courtesy as of valour; so I shall borrow him as a type of Romanée Conti, ranking the others as his brothers-in-arms.

For the suavity of these wines would not show to such perfection were it not for their valour and sturdiness. They are indeed princely wines, and the qualities that lurk in their depths are courage and fair dealing.

And what shall we say of Richebourg: Richebourg that has given me the most recent of my Three Great Bottles? It was far from the first occasion on which I had blessed the noble name; there was a Richebourg 1915 that I met in 1935 which caused me to think that twenty years was the ideal age of Burgundy; and there was another bottle, this time a 1911, that Guy Knowles had poured for me a year or two before, when it broke the lances of four other first growths, some of them of years that are usually set far above 1911. But the Bayard of Richebourgs was a 1923, drunk at the invitation of Sir Francis Colchester-Wemyss at his club on the night of Shrove Tuesday this year; and I have just had to refuse an invitation to the same hospitable table, when I feel sure another of these precious bottles is going to be drunk. Such are the horrors of war! For this, indeed, was a worthy rival to the La Tache 1904 of which I have just been writing. Its colour arrested my attention at once. Claret can show the clearest limpidity; but a truly limpid Burgundy seems to have more substance in the colour; just as translucent, one feels that it would take more than a mere pebble to crack *that* glass. And this Richebourg had the true glow of the ruby; you were conscious of a lapidary worth. The bouquet did not fail to justify the clarity; it was almost spiced, so sweetly aromatic was it. And then the true glory revealed itself, not to the eye, not to the nose, but to the palate. It caressed the gullet; it spread its greeting over all the mouth, until the impatient throat accused the tongue of unfair delay. It was glorious, glorious, glorious; and a month later I had not yet stopped talking about it. For I had forgotten how good Burgundy could be. It was fourteen years since the La Tache; and, although there had been many fine bottles in between, none of them had been of that rare outstanding merit that the Three Great Bottles reached. A grateful heart may be allowed a long chant of thanksgiving. [CAMPBELL: *Maurice* in excelsis; *the real Maurice and the right Maurice, The* 1923 *Burgundies are no doubt the best to be found to-day.*]

There are two other first growths in Vosne, neither of which ever reaches an English cellar; Les Verouilles, extending to about eight acres, and Les Malconsorts, comprising about fifteen. But what I have not infrequently seen in our wine-lists is a Vosne-Romanée ranked as a prince of the blood, sometimes priced higher than a La Tache or a Richebourg of the same year. Now, that particular merchant or restaurant may be acting with complete good faith; having bought from a Burgundy shipper parcels so labelled, he may have found that the wine with the mere district name has proved a better Burgundy than the more aristocratically labelled one. But, if they both come from the same shipper, the shipper cannot have been playing the game; for the wine from a first-class vineyard must always be better than a mere *ordinaire*. It is an extraordinary fact that this particular eccentricity of labelling succeeds in nine cases out of ten; and the purchaser never queries a nomenclature that ought to put him on his guard at once.

We now reach Nuits, the last commune of the Côte that bears its name; and here, of course, the principal vineyard is Les Saint-Georges, the borrowing of which name started all the trouble. This vineyard is about twenty acres in extent, and is bounded by Les Vaucrains (about fifteen acres) and Les Cailles (ten acres); the map shows a *clos* called Les Didiers to the immediate south, but although this is coloured as a First Growth, it is not set forth in the list, which, however, also includes Les Pruliers (thirty acres), Les Porrets (about seventeen acres) and Aux Murgers (twelve acres). Of these, the first three are the only ones I have met; and the Saint Georges not often in security. Once a club to which I belonged asked its merchants to send samples of various types of Burgundy, including what they described as Nuits-Saint-Georges. One firm sent up a bottle of a beautiful Les Vaucrains; it was sent back with a note to say that what was wanted was a Nuits-Saint-Georges! Any of these first growths may be relied upon to take care of the credit of Burgundy. But it is significant that, although there are twenty or thirty named vineyards in the district that are graded as first, second and third growths, there is not a name amongst them beyond those I have already mentioned that could be recognized by those familiar with our wine-lists. They have all taken refuge under the banner of Nuits-Saint-Georges.

We now come to the Côte de Beaune; and here we find only one great vineyard: Corton; although the Clos du Roi is often reckoned as almost as distinguished. On the other hand, there are a far larger number of lesser vineyards that have made some publicity for themselves, such as Les Bressandes; and they have a habit of repeating themselves, so far as nomenclature is concerned, in successive communes. But the most remarkable feature of the Côte de Beaune is the way that the Clos du Roi has ministered to the fame of its rival, Corton. When the communes began the bad habit of attaching the name of the most famous vineyard to the name of the parish, Aloxe became Aloxe-Corton. So far fair enough; but one would expect that if the Clos du Roi had wanted to put the name of the commune before its own, it would have called itself Aloxe: Clos du Roi. Instead of which, it has become Corton: Clos du Roi.

In the parishes of Savigny and Pernand which adjoin Aloxe-Corton, there are three vineyards with names that contain a word not unfamiliar on English wine-lists. These are the Ile des Vergelesses and Les Basses Vergelesses in Pernand, and Les Vergelesses in Savigny. It must be remembered that it is the wines of Savigny that, according to a very old inscription, are 'nourissans, théologiques et morbifuges.' The first and last qualities are apparent; the middle one may explain my frequent invasion of the sphere of religious controversy. The three Vergelesses vineyards, which adjoin one another, can be guaranteed to acquit themselves in any competition, whether lay or ecclesiastical. In Savigny there is also a vineyard of repute called Les Marconnets; but whenever I have had the good fortune to drink a bottle so labelled (and it was usually a very good bottle), I never was sufficiently informed as to whether this was the Savigny wine of that name or a wine from the vineyard in Beaune that bears the same designation and is ranked on the same high level.

For Beaune, being the capital of the wine-district, seems to have adopted a seignorial right of assuming to its own vineyards names that are to be met in the neighbouring districts as well. Here, for example, we find not only Les

Marconnets competing with its rival in Savigny, but also Les Bressandes, not to be confused with the Aloxe vineyard of Les Bressandes, and Les Grèves, also apt to get mixed with its namesake in Aloxe. There is also Les Cras, repeated as a name somewhere in the neighbourhood, and the confusing pair, Clos de la Mousse and Clos des Mouches, both of which, with the others I have mentioned, are counted first growths of Beaune. You rarely find them over here. But Beaune is remarkable, not only for its own wines, but also for the wines given as an endowment of its Hospice. A wine labelled 'Beaune Hospice' need not have been grown in the Beaune commune. The Hospice is one of the most wonderful charitable monuments in the world. It was founded for the benefit of the sick poor in 1441, for that is both the date when the Chancellor, Nicholas Rollin, acquired the site, and the date of the Bull of Pope Eugene IV, by which the foundation was decreed to be free from all ecclesiastical jurisdiction save that of the Pope himself: a condition that lasted until Napoleon's Concordat of 1801.

This foundation was served by pious ladies who embraced the same rule as the *béguines* of Belgium. They were not unpaid; they each enjoyed a salary of three francs a year! One cannot imagine the sourest reformer accusing them of waxing fat on that allowance. Fortunately, at an early stage it became fashionable to endow the Hospice with vineyards to provide the means of carrying on its good work; and the annual sale of the Hospice wines, which takes place in the convent itself, and is conducted on the old sale-by-candle system, has a marked influence on the other Burgundy markets, and often tends to govern the prices for the year. Some of their wines, notably the Enfant Jésus, are well known in England, and on the whole I have been lucky with the label 'Beaune Hospice.'

We next reach Pommard; and it cannot be too often repeated that Burgundy is not to be summed up in 'a bottle of Beaune or a bottle of Pommard.' For Pommard is a commune of the Côte de Beaune, and its wine is properly called a Beaune, as distinct from a Nuits. I admit that this would not be very accurate terminology; but the true contrast lies between the Côte de Beaune and the Côte de Nuits, not between Beaune and Pommard. Of course, your 'bottle of Beaune or bottle of Pommard' is just what you deserve to get for asking for it like that; it will probably have come from Algeria, and will have paid many commissions before it is shaken up and put on your table. But a good and genuine Pommard *ordinaire* (and the same applies to Beaune) should be a limpid, clear and full-flavoured wine, entirely free from that horrid taste of the sugar-boiler that proclaims the bogus upstart at once.

The principal vineyard is Les Epenots, which, with Les Rugiens-Bas and Le Clos-Blanc, forms the Pommard contribution to first growths. Amongst the less highly-classified wines must be mentioned the Clos de la Commareine, which has always enjoyed a great reputation in this country. It is a long time since I last saw it on a wine-list, but it has left fragrant memories behind it. And I must also mention a vineyard which I cannot find in the classified list of Burgundies, but to which I owe many a pleasant evening: Les Epenottes, shipped by Louis Poirier. Indeed, the Les Epenottes of 1926 was my first introduction to a light Burgundy bottled as it was made; and at seven years old it was a revelation. It had the brilliant clearness of a good red ink; it had no sediment whatsoever; and it had a perfume akin to that of cedar-wood. It was a beautiful wine; and it well matched the same shipper's Montrachet Bâtard, which I have drunk in at least two vintages at the Hind's Head, Bray, and cannot praise too highly. I think that both the wines in question have gained from the policy of Monsieur Poirier, who apparently insists that a wine should not be 'made up' at all. Now, in what way are you meant to translate that sinister phrase? In this. Most Burgundy vineyards are small; Les Epenottes, I believe, only produces about the equivalent of seventy dozen bottles a year. Now, a wine-merchant in a large way of business that desired to put a particular wine on its list would probably want the whole of that seventy dozen to enable him to show fair play between his individual customers. But the agent would also want to show fair play between *his* customers; and the shipper would want to show fair play between his; so that the upshot would be that the owner of Les Epenottes would find it very hard in the ordinary way to persuade a shipper to buy his wine and sell it under its own name. The shipper would probably say, 'Yes, I will buy your wine, but I shall use it to blend my "Pommard" or my "Beaune Supérieure" of this year'; and the owner would have to allow his wine to be blended in this fashion with the wine of a lot of other small vineyards, so as to make a thousand dozen or so of a standard wine, or else to go without a purchaser. Monsieur Poirier apparently made up his mind that there might be a few individuals who would club together to take the whole of the seventy dozen and divide it amongst themselves; and by the kindness of André Simon I benefited. The result was most gratifying, and I hope that the Burgundy shippers will in future give a chance to the gourmets to take amongst themselves the whole output of every little vineyard that can produce such a distinctive wine.

So we pass from Pommard to Volnay. Many years ago my good friend, John F. Bourke, told me that whenever he was in doubt as to what to order by way of a bottle of Burgundy he always asked for a Volnay, 'for,' he said, 'it is not as widely known a name as Beaune or Pommard and thus is not worth the faker's while.' This has often proved to be good advice; but I am not guaranteeing that every bottle labelled 'Volnay' can be accepted implicitly as genuine. In the first place, it is only a district label; and unless there is added the name of one of its famous vineyards it is only an *ordinaire*. And it has four first-class vineyards: Les Fremiets, Les Champans, Les Caillerets and Les Angles; and of these I cannot believe that the others would deny precedence to Les Caillerets. When I first joined the Windham Club in 1922 there were still a few bottles of Les Caillerets 1889 in the cellar. I drank one bottle with a good friend, and it was my first introduction to great wine. Now let me protest at once about the ordinary method of approach to a great wine. The average host bids you to dinner, and, without a word of warning, sets you drinking Lafite 1864, or something like that. The result is that he is robbing you and the wine of a great deal of your due. If you are going to drink a great wine, you should be told that you are going to drink a great wine. You should have an opportunity to prepare your palate and get it into a proper state of receptivity, and your mind into a proper state of reverence.

You should drink barley water for lunch, and go without your tea; you should avoid the barber's shop and the Turkish bath, lest the first might spill his highly perfumed washes upon you, or the second create an unnatural dryness of the palate. You should go through the preliminary stages of your dinner with your mind fixed on the great moment, and you should not allow the conversation to be directed into controversial channels. Later, yes, but not until the wine has first had your whole attention. And you certainly ought not to fail to call to mind the goodness of God that has given you such a gift; and at least in this very reverent sense you should drink it in remembrance.

Now my good friend—why should I not name him? it was Michael Eddis—performed his duties like a true host. He began talking about the Volnay-Caillerets about a month before the dinner, and he drew such a picture of its marvellous greatness that I was all agog to drink it. He ordered it several days before, so that it had a good opportunity to stand up and let all the sediment get to the bottom of the punt, while it gradually took the temperature of the dining-room. Then he had it carefully decanted just before dinner; and he ordered the rest of the dinner so as to produce the gradual crescendo and avoid the anti-climax. And so the moment arrived when it was proper for me to raise my glass. This was nearly twenty years ago; but I still remember the magnificent shock of that bouquet, rich in mellow perfection, and entirely free from the infirmities of age. I took one sip; I closed my eyes, and every beautiful thing that I had ever known crowded into my memory. In the old fairy tales the prince drinks a magic potion, or looks into a magic crystal, and all the secrets of the earth are revealed to him. I have experienced that miracle. The song of armies sweeping into battle, the roar of the waves upon a rocky shore, the glint of sunshine after rain on the leaves of a forest, the depths of the church organ, the voices of children singing hymns, all these and a hundred other things seemed to be blended into one magnificence. I have since met two other bottles worthy to be compared with that one, and they were both bottles of Burgundy. Yes, I, a devotee of Bordeaux, solemnly declare that the three greatest bottles I have ever tasted were all from Burgundy, the others being the La Tache 1904 and the Richebourg 1923. But the Caillerets 1889 was the greatest of them all. [CAMPBELL: *I think this proof that Maurice was really a Burgundy rather than a Claret 'fan'! He almost admits as much in the next chapter when he accords 'most grateful thanks' to a* 1911 *Moulin-a-Vent, 'in a bottle of the Burgundy shape,' because it gave him a taste for red wine. He was too strict a purist to call it Burgundy though it appears as such on most wine merchants' lists, as he must have known.*]

After Volnay we say farewell to the great red wines, for Monthelie can be agreeable, but is never great; and Auxey has nothing wonderful to show us. We then reach the white wine country, and it may be a surprise to some to learn that Meursault produces one good red wine, Les Santenots-du-Milieu, while Chassagne has the Clos Saint-Jean and La Boudriotte, and Santenay gives us Les Gravières. I have not met any of these last myself.

This brief excursion brings us to the end of Burgundy proper, and Beaujolais deserves the compliment of another chapter.

CHALONNAIS, MACONNAIS AND BEAUJOLAIS

'THE principal red wines of Burgundy,' wrote Mr. James Richmond Sheen in 1864, 'are Romanée Conti, Clos Vougeot, Chambertin, Corton, Volnay, Pommard, Beaune, Macon, Beaujolais, etc.' Seventy-six years later, English *restaurateurs* and even some wine-merchants continue to classify Beaujolais amongst the Burgundies; but as it rubs shoulders on such occasions with Côte Rôtie, Châteauneuf du Pape, and even Hermitage, its own inclusion must be considered a venial sin. Burgundy proper stops at Santenay; the Côte Chalonnaise succeeds it, while the wines of Macon are further to the south. Then only do we reach the Beaujolais country, the wines of which are distinctive and delicate, and of a character completely different to the Burgundy wines. Before dealing with them, let us just glance at the districts to the immediate north of them, where wines deserving of more than a passing mention are made.

But we are at once in difficulty about nomenclature. Of the Chalonnais, the two best-known wines pass under the names of two parishes: Mercurey and Givry. I have never seen a bottle of Mercurey that has condescended to localize itself more particularly; yet a bottle of Mercurey means no more than a bottle of Beaune, or a bottle of Volnay; it is a district, not a vineyard, wine. According to Mr. Morton Shand, the best *clos* of Mercurey does not come from the parish of Mercurey at all, but from Touches, a parish next door; it nevertheless calls itself Mercurey: Clos du Roi, imitating its great namesake at Aloxe. A bottle labelled 'Mercurey,' if of good provenance, should taste nearly as good as a Beaune or a Volnay, and will probably be cheaper. We are still in the country of the Pinot, and all good Mercureys (and good

Chalonnais wines) are made from the noble plant, the Gamay having refrained from claiming royal honours in this district. What has been said of Mercurey is, I believe, also true of Givry, although I have never met that label over here. Nor have I come across the white wines of Rully which come from this district, and are supposed to be the competitors of those of Pouilly

Macon at once suggests Lamartine; but it can hardly produce a wine, or at any rate a red wine, worthy of him. The truth is that Maconnais and Beaujolais wines are almost impossible to classify apart; Romanèche Thorins is on the borders of Beaujolais, so that the famous wine of Moulin-à-Vent is claimed by both districts, although more frequently adjudged to Beaujolais. But in white wines Pouilly does honour to the name of Macon.

In Poperinghe in 1916 the best-known of the war restaurants was in the Square; it was called 'La Poupée,' and a mother and her unlimited family of daughters served meals of which the best Paris houses need not have been ashamed. The youngest girl cannot have been more than ten years old, but she joked with the officers with a readiness of repartee that would have done credit to a young lady of twice her age. Some years ago I saw a letter in *The Times* stating that this war-baby was getting married, and asking that those who still remembered 'Ginger' would contribute to a wedding present. I believe that the result was enormous. I hope that poor Ginger is not caught up in the present war as well.

But it was not about Ginger that I meant to write, but about the wine we drank at 'La Poupée.' Remember, I was to all intents and purposes a teetotaller in those days; but I had learned in the trenches how a glass of Port could restore a chilled circulation, and I was not averse to try a few experiments in pleasanter surroundings. So at the dinners that we used to take when out of the line we used to bid Madame to bring us a bottle of 'Courage.' And she would place upon the table a bottle of the Burgundy shape, bearing the label 'Moulin-à-Vent, 1911.' You will observe that the wine was five years old, which is a good age at which to drink Beaujolais. They mature their wine in casks rather longer than in Burgundy proper or Bordeaux, sometimes not bottling until five years after the vintage. This particular wine must have been a good while in bottle, for it had grown to like its new home, and was entirely free from the vices of new-bottled wine. It had a beautiful colour, not so light as to suggest very late bottling, but at the same time lighter than the usual Claret or Burgundy. It had a very full bouquet, rather of the raspberry variety. It was a balanced wine; it appealed to every sense, and you could drink a couple of bottles of it without being aware that it was having any effect upon you; I will not answer for what the neighbours thought. But to that Moulin-à-Vent 1911 I owe my most grateful thanks; not only because of the comfort that it then gave me, but because it was that wine that taught me to like the taste and bouquet of red wine. Claret followed, but much later. I did not then know, nor indeed do I know with certainty now, but I believe that Moulin-à-Vent is the name of a district in Romanèche-Thorins, and there are several separate vineyards that are entitled to use the name, each of them having a distinguishing label of its own which I have never seen, nor do I know any of their names. Here one must trust that one's wine-merchant has bought from a reliable Beaujolais shipper; I venture to guess that the separate vineyards are each on the small side, and that they submit to blending, like the small Burgundians.

The most interesting feature of the Beaujolais wines is that they are made from a Gamay grape. It is not the Gamay that is so reprehended in Burgundy; it is a cousin, that apparently is free from the vices responsible for its relative's unpopularity. For a good Beaujolais is a wine calling for no apology. Lighter than a Burgundy, it is of about the same 'weight' as a Claret, but the flavour is entirely on the Burgundy side. Burgundy is always a man's wine; Beaujolais has a more feminine grace. Burgundy graces the table of the Guards; no Light Infantryman need be ashamed of Beaujolais. It is a sweet wine, not in the sense that a Sauternes is sweet; but there is a seductive charm about it, and a freedom from asperity, that makes it most attractive on all occasions when a light wine is appropriate. It is tasted best against a dish that has had just the right acquaintance with an onion. In Lyons it can be appreciated against the perfect background and in its own perfection. I should not hesitate when a visitor to those parts to prelude liberally on Beaujolais and to follow with a fugue of Burgundies. The beds in Lyons are ample and convenient. And breakfast is always an agreeable meal in that city of Jehu, for neither Beaujolais nor Burgundy bears malice.

The other leading wines of this part of the Beaujolais (*pace* Macon, which appears to me to have an equal claim to their honour) are Fleurie, Morgon, Chénas and Juliénas. Each of these names covers a number of vineyards; but the vineyard label is never seen over here, and hardly ever does one even meet the name of the district. I believe that I did once come across a bottle of Les Carquelins, this being one of the best Moulin-à-Vent vineyards; but I would not pledge my word upon it. And although I have frequently asked for a little Moulin-à-Vent for old times' sake, I have never yet succeeded in finding any about the pedigree of which I could get the assurances I desired.

The fact is that the word 'Beaujolais' is one of a series used in this country quite recklessly, and with very little precaution taken to ascertain the exact *provenance* of the wine. A bottle of Beaujolais on the average wine-list means a bottle of wine of a Burgundy type that is a little light on the palate. Years ago I used to cellar a cheap Beaujolais that made pleasant drinking on a winter afternoon; the wine-merchant was a dear personal friend who used to say the most startling things about his own stock, and he always insisted that this particular wine was a blend of his own, softened with glycerine. I do not take him, even now, quite seriously; but I am quite sure that the wine so labelled had never seen the vineyards of Beaujolais. Yet he was selling large quantities of it as a stock wine that could be indefinitely repeated; and he told me that he never had a criticism of it. This shows in my opinion that the average drinker of wine in this country not merely knows nothing about it, but has only the vaguest ideas on the different tastes and bouquets of the various types of wine. And I, a fervent upholder of the professional wine-merchant, make this one criticism of that excellent person. I think that he should do more than he bothers to do to-day to educate his average customer. I know that I am always a welcome guest at a number of hospitable tables in London and the provinces, and that I shall never ask in vain for the assistance of any wine-merchant in a point of instruction. [CAMPBELL: *Maurice is right: he was a welcome guest at any wine-merchant's table in spite of the fact that he had been a teetotaller till he was thirty (he died before he was sixty). He was a lively and amusing talker, an apt pupil, he liked wine and strove to learn all about it, and it was well worth going out to teach him. Would there were many others like him. To give one of his own Biblical illustrations he was like the prodigal son who came to his senses and to his father's house and was treated to the fatted calf and, no doubt, the best suitable liquid refreshment. He soon began to teach us in the wine trade our business; readers of this book will perceive that that was really his métier and he was a witty, persuasive and pig-headed preceptor.*] But he will allow Mrs. Smith to go on ordering her monthly dozen of Saint Emilion 1931, or Beaune Supérieure 1922, without ever trying to teach her how to use her own judgment in choosing a wine, thereby denying her one of the most amusing of a housekeeper's tasks. Probably he will defend himself by protesting that no woman could ever learn to pick a wine for herself. 'No woman' is too harsh a phrase; I know at least a half-dozen to whose judgment I should defer, including a certain lady not a thousand miles from the Hind's Head at Bray, and another in Dean Street. But I have generally found that a woman's palate is more to be trusted in the matter of food, especially sauces; there the mere man tends to clumsiness—better at drams than at scruples, perhaps. The right touch of tarragon, the judicious blessing of lemon—above all, the *pianissimo* of garlic; these are usually feminine arts. But in the choice of a Claret or a Burgundy, to say nothing of Port or Sherry, give me a man's decision every time.

Now, who says that Beaujolais cannot start a conversation?

THE RHÔNE

LYONS is a city of good living. It is the city of silk; not merely in stuffs but in stuffings, for I can hardly imagine a place where the cloyed palate could be more seductively won back to an admission that there still might be dishes in the world worth the tasting. The wine of Lyons is Beaujolais; but at the same time it recognizes that there are greater heights and depths than the good Gamay can reach, and it does not hesitate to stock its cellars from the nobler regions further north and it even extends a tolerance to Bordeaux. But south of Lyons there commences a wine country that must always be held in reverence by the more experienced wine-drinkers. Tourists may stand on a bridge in the city and witness the unwilling union of the Rhône and the Saône; the one still bearing in its waters the icy glint of the glaciers, the other scorning to colour its depths with any hue borrowed from the heights. A few miles below the city the union is complete, and the ice-water of the Alps has mingled with the warmer current from the Côte d'Or so as to be indiscernible. The combined flood sweeps southward to the Mediterranean, flanked by vineyards that are perhaps the most constant and reliable of all the French wine-lands; for the climate is warm and unvarying, and the visitations of hail and frost appear to be rarer. On the other hand, this freedom from the accidents that frequently afflict Bordeaux and Burgundy is paid for by an absence of the *finesse* that so often distinguishes the wines of those districts. Rhône wines are rugged wines. They can be magnificent, as in the case of Hermitage; but I prefer to consider them in the light of good companions rather than as grave counsellors or inspired poets. [CAMPBELL: *Rhône's retort might be* tu quoque.] For an autumn luncheon I cannot imagine a more agreeable accompaniment than a Côte Rôtie; but I would not serve it at a dinner meant to be an honourable feast.

Who was it that were accustomed to lunch *al fresco*, producing from their game-bags 'un bon morceau de bœuf en daube, des oignons crus, un *saucissot*, quelques anchois, et commencent un déjeuner interminable, arrosé d'un de ces jolis vins du Rhône qui font rire et qui font chanter'? Why, who but Tartarin de Tarascon and his friends, engaged upon 'la chasse de casquettes'! Yes, the wines of the Rhône may lack their Cluny, their Lamartine, their Montesquieu, their Vergniaud; but bright in their crown shines the jewel of the praise of Tartarin, and the fact that they furnished the drinking of that remarkable man is surely glory enough. And if he found that these wines made him laugh and sing, who are we to detract from that joyous tribute? At the same time, *autres temps, autres mœurs;* and those amongst us who are setting out on a gay enterprise will probably lift our hearts with a different glass. For I confess that I find the Rhône wines lacking in a sense of humour; they are grave and serious in their approach; they would scorn the lighthearted *insouciance* of a bottle from Anjou or Touraine, and their disdain towards a Mosel would not be merely because it was a German. They would consider Champagne and its lesser imitators a fuss about nothing; they would probably regard Burgundy or Bordeaux as a worthy yeoman would look upon the nobility, persons to be respected but not envied. France has given us many types of mankind, but in the long run she is best exemplified by the peasant. For that reason, although those who search for the finest French wines will always turn to Bordeaux or Burgundy or Champagne, it is perhaps true that the wine that best typifies the French character in its simplicity and sincerity and rugged obstinacy is the kind of wine that comes from the vineyards along the course of the Rhône.

The three names from the Rhône that have become familiar to those who drink wine in this country are Côte Rôtie, Hermitage and Châteauneuf du Pape. They have in common with all Rhône wines a curious touch of flavour that suggests the slight smoky acridity you can experience if you strike a couple of flints together under your nose; it is usually called the 'gunflint' flavour, and it is not in the least disagreeable: quite the contrary. In my experience of these wines they have always shared another quality. There is a touch of rust in their redness. It is not the *pélure de l'oignon* that so often delights us in an old Claret; it does not suggest age, and on the other hand it does not detract from limpidity. The red is a slightly browner red, that is all. Whether this is the effect of the gunflint upon my imagination I cannot say. Look for yourself when next you drink them.

Côte Rôtie comes from Ampuis, a few miles south of Vienne and about twenty-five miles from Lyons. The Syrah grape, supposed to have been brought back to Hermitage by a crusader and thence spread along the whole course of the river as far as Lyons, has here modified its name again (for it was originally called Shiraz, from the Persian city of that name, Persia being its supposed place of origin) to Sérine; another vine of the locality is called Persan, as though determined to proclaim its Iranian origin. The wine is deep in colour and full of body; a wine to chew, as some people say. I should never speak of it as being a fine wine, but it has a certain breeding, and a good sample proclaims itself an honest product of the vine in no uncertain manner. Unfortunately, Côte Rôtie is another of these names used carelessly by those whose duty it is to know better; and in the cheap wine-shops of Soho it is quite a common label, used, alas! to conceal dreadful and entirely bogus liquor. But a bottle of good provenance will prove a good friend, and comfort a heart bowed down by weight of woe. Some find a touch of raspberry in the flavour; I can see what they mean, but it is the fruitiness rather than the sweetness that is borrowed. I am told that there are only five genuine Côte Rôtie vineyards; if so, the disappointment behind that Soho label is more easily understood.

The white opposite number to Côte Rôtie is Château Grillet; and it is worth while noting that although the slope is roasted the château is not grilled. I see that Mr. Shand pulls the august leg of Mr. Warner Allen for having spelled the name 'Château Grillé,' but many others have made the blunder, if blunder be not too harsh a word, being excused or at least extenuated by the fact that the wine is very rarely found over here. Oddly enough, it was Mr. Shand's own hospitality that first introduced me to it, the particular bottle having attained an age of between sixty and seventy years. I think the vintage was 1861, but I am not sure; I only remember that it had preserved its vinosity, and the clean, pebbly taste was quite remarkable. It seemed to leap the intervening hundred miles or more to link hands with Chablis; but it appeared to me to be of a majesty greater than that attained by Chablis, or any Burgundy except Montrachet. Its rarity prevents Château Grillet from being much imitated. The vineyard is situated at Saint-Michel-de-Condrieu.

Going further down the river, we find at Tain, on the left bank, the famous Hermitage district. In pre-phylloxera days Hermitage could challenge almost any wine as regards staying power; but, like Bordeaux, it has grown shorter-lived. But of all the clanking warriors that ever strode into a dining-hall, commend me to a really fine old Hermitage. Whenever I hear the fourth movement of Beethoven's Eroica Symphony I see the booted and spurred cuirassiers of Napoleon's army trying to tread a measure with the vivacious ladies of Vienna. It is of such warriors that my Hermitage reminds me; but I think that the wine would more rapidly grow impatient of the dance and break out into the fierce Hungarian march that makes the interlude. For it is the manliest wine that can be drunk. At the back of its suavity and fruitiness is a touch of something approaching to acerbity that corrects the impression of gentility and makes it clear that this is no carpet knight or Whitehall warrior. By the way, as Mr. Shand pulled Mr. Warner Allen's leg about the Château 'Grillé,' shall I in turn let a cat out of the bag and expose his quotation from Professor Saintsbury, in which he makes the latter speak of Hermitage as 'the *manifest*[1] of wines'? I am not sure that Mr. Shand has not made a happy error, and that the manifest of wines would not be a good sobriquet for Hermitage, which is indeed the virtue and courage of wine made manifest. White Hermitage I hardly ever have met; but the red can be magnificent. There are not a large number of individual vineyards; but when the former proprietor of the Festival Theatre at Cambridge went out to settle at Hermitage he purchased a Clos Saint Patrice which is said to owe its name to the fact that the National Apostle of Ireland halted there on the way to his task of the conversion of Ireland (which some people are mean enough to suggest that he left incomplete).

[1] Professor Saintsbury wrote 'manliest.'

This is a pretty legend, and, having drunk the wine, I gladly recognize that the vineyard had a blessing from some source or another. But Saint Patrick is supposed to have landed in Ireland in the year 432; and the pious crusader who brought the Syrah grape back with him from the East is not thought to have arrived at Hermitage until the first half of the thirteenth century. No doubt there was a vineyard there in Saint Patrick's day; but it did not produce Hermitage, as we understand the word.

A little further south, and back on the right bank of the Rhône, we find Ardèche, in which district good red wines are found at Tournon and Cornas, and good white wines at Crussot and Saint Péray. The reputation of the white wines exceeds that of the red; perhaps because of the popularity of a sparkling white wine made at Saint Péray. But I have drunk very agreeable bottles of red wine labelled Ardèche; the quality is akin to red Hermitage, and there is still the stone-like hard cleanness that is such a feature of the wines of these parts.

We have to travel southward for nearly another hundred miles before we reach another famous name. Vaucluse suggests love rather than wine; but a name that recalls Petrarch has given itself to a district that includes the commune of Châteauneuf du Pape, a title that goes back to the days when the true Pope reigned at Avignon while a usurper reigned at Rome; when Bertrand de Goth was called from Bordeaux to become Pope Clement the Fifth, giving his name to glory by thus christening his Bordeaux vineyard, and to infamy by lending himself to the shameful plot of Philippe le Bel against the Knights Templar. Pope Clement found at Avignon much to comfort him for the loss of his red Graves; his summer palace contained vineyards that produced a very distinctive wine, and the name of the palace has survived as the name of the wine that used to be made there. The best-known vineyards there to-day are the Château Condorcet and the Château de la Nerthe; I have been unable to discover any connection or analogy between Château Condorcet and the philosopher of that name save that of merit and distinction. It will, however, be remembered that Condorcet joined the Girondins!

The wines of Châteauneuf du Pape are full-bodied and

expansive. They have a beautiful colour, albeit still with that rusty tint; and they develop a very full bouquet. A local saying proclaims that they spread their flavour like a peacock expanding his tail in your mouth. They do not lack their poet, either, for Frédéric Mistral has sung in their praise. As an example of the way districts can vary in their quality in a given year, while agreeing with Burgundy in calling 1906, 1908, 1911 and 1919 good years, they also praise 1916, 1918, 1920 and 1922, years rather contemned further north, although Hermitage and Côte Rôtie would probably not disagree, having also the benefit of the more constant hot summers.

Tavel, a little spot about ten miles north-west of Avignon, makes a wine, light in colour and light in body, that is a delight to the eye as it is to the palate. For those occasions when a light wine is sought I can imagine no more agreeable candidate. I have no hesitation about drinking it with fish, and it consorts just as readily with meat dishes or even game, though for such purposes I do not strongly recommend it. When I say that it was beloved by Brillat Savarin in the one world and by Balzac in the other have I not set it up in both gastronomy and in letters?

So ends the wine country of the Rhône, and with it all except one of the important wine districts of France, for the Midi, producing any amount of table wines, does not run to distinction. The exception is the Jura, where the wines of Arbois, red and white, are worthy of the attention of the connoisseur. They have an affinity with Sherry; they are capable of attaining unbelievable age without any fortification; and they have an agreeable and distinctive flavour, which in the case of the red wines is something apart. The late Monsieur Jean Sender, whose tragic death was a great loss to London's wine-lovers, introduced the red wine of Arbois *en carafe* when he opened his restaurant 'A l'Ecu de France.' It did not pretend to be anything more than a table wine, but it had the authentic quality, and at the price at which it was sold was a very good bargain. Finer wines from Arbois are not often met in England; I think that Philip Trotter is the only wine-merchant that has specialized in them. They repay a little attention.

PORT

IS Port to be called a wine? For wine is supposed to be the juice of the grape, fermented, but not otherwise treated. And Port owes its quality and merit to the fact that it is not a natural fermented grape-juice; it is the produce of the grape and nothing else, but the fermentation of the juice has been checked by added brandy, and the flavour has been doctored by the addition of a condensed wine, called jeropiga, which differs in the house of each shipper to such an extent that the expert rarely doubts whether he is drinking A's or B's or C's shipping of a particular vintage. The best piece of invective in modern English is to be found in Mr. Morton Shand's *A Book of Wine*, the first two pages of the chapter on Port being calculated to frighten off any neophyte and make him vow to stick to Claret or Burgundy for the rest of his life. On jeropiga Mr. Shand is particularly alarming; he says that the most popular formula includes dried elderberries, unfermented must, black treacle and a spirit distilled from figs. I have found Mr. Shand so helpful about other wines that I have felt it to be my duty to give attention to these suggestions of his about Port; and I am bound to say that, if one may swear by the card, he might be able to make out a good case. But the man who calls for a glass of Cockburn or Taylor or Graham or Croft or Dow or Warre or Fonseca or a dozen other familiar names can be quite sure that the wine he is drinking would never come under the maledictive influence of Mr. Shand's description. [CAMPBELL: *I am sure the half-dozen or so shippers whose names are mentioned will feel relief and gratitude, but they are a long way from being the only pebbles on the beach. If a tithe of the foolish and false rubbish that has been written in books about Port were true, the wine would have disappeared from our tables as rapidly and completely as the said books have.*] Even the logwood with which he suggests that colour is sometimes corrected knows no place in the lodges of reputable firms. 'Red Biddy,' no doubt, is generally sold as Port; but I have always understood that methylated spirit plays a prominent part in the manufacture of that communist Irishwoman, and any other ingredient after that must be a venial sin. I suppose that there are houses of call which, relying on their beer for their principal trade, feel that they must stock something labelled Port for the benefit of their more 'refeened' lady customers; and they are not very particular as to the history of the bottle so labelled. It is true that by law nothing may be sold as Port save wine made from grapes grown in the valley of the Douro and shipped over the bar of Oporto. But there are more ways of killing a dog than choking it with a pound of butter; and the rarity of prosecutions for fraud on the name of Port, coupled with the universal contempt for what is popularly called 'public-house port,' leads me to believe that there is a considerable quantity of wine sold in this country under the name of Port that would merit Mr. Shand's most savage apostrophe. As regards the genuine article, Doctor J. L. W. Thudichum visited Oporto in 1893 and made a most elaborate study of the process of manufacture. He says:

> 'It is said that port wine is coloured with elderberries and other dyes, and sweetened with jeropiga and treacle, besides being dosed with brandy; but I have been unable to find any evidence of this, at least as regards Alto Douro wines. . . . Logwood is never used in trade for dyeing anything purple . . . and although it may occur that particular artists in mixing and counterfeiting dye some pipes of white wine with elderberries and give them astringency with logwood, nevertheless I believe that such products would commercially not pay the cost of their production. . . . I am therefore of opinion that the sooner we dismiss these prejudices and errors regarding elderberry and logwood in port wine the better.'

He corrects his remarks, so far as jeropiga is concerned, a page or two later by adding: 'If the wine is not sweet enough a quantity of jeropiga is added; this, when legitimate, consists of sweet must preserved by the addition of one-fifth of its volume of brandy. . . .' I ask to be corrected if I err in my opinion that each shipper has his own distinctive recipe for jeropiga; I cannot otherwise understand how the shippings of the same house in vintage after vintage can repeat the flavour that identifies their produce.

Let us therefore start by saying that there are Ports and 'Ports.' And, as this book is intended to be read by intelligent persons, it seems to me that we may eliminate from our consideration all except the wines shipped by reputable houses, wherein we shall not suspect the presence of elderberries or logwood or methylated spirit or Californian Spirit of Figs (not to be confused with a popular medicine of unimpeachable character) or any other deleterious ingredient. And perhaps, having adumbrated the existence of libels and slanders on the character of Port, it will be worth our while to begin at the beginning and tell the story of the wine, both historically and with reference to its own criteria.

Port is Oporto wine; that maritime, or at least seafaring, city has always been the gateway through which the wine named after it has reached other countries. But the wine so shipped was not always what we to-day call Port. Most of it was made from the grapes grown along the banks of the Upper Douro; but there were other, lighter wines that came in from other districts which would have been graded as light as Claret but which lacked the elegance of the French wine. The Douro flows through a series of rocky valleys, the sides of which have been terraced in the stone; and in these wild regions have been cultivated the grapes of a vine far richer, juicier and sweeter than anything grown in Bordeaux or elsewhere. To be accurate, there are several species of vine: seven, if my memory serves me; but there are five or six in Bordeaux, and yet the Cabernet is named as the informing vine of Claret. So in the Douro the various grapes are alike in being of a more luscious variety, although I believe they taste a bit acid to the tongue when plucked from the vine.

When gathered, the grapes are thrown into a stone *lagar* or trough, about two feet deep and anything from ten to thirty feet square. Here they are trodden by the vintagers; and it is not usually appreciated that fermentation can be hastened or retarded by the number of men put to the labour, the heat of their bodies helping on the working of the yeasts. The first treadings merely pulp the grapes and extract the juice, the later pressings extract the colouring matter from the skins. The test of the completeness of the extraction is to pour a little of the juice across a white plate, and if it leaves behind a streaky red stain that is considered the mark of success. The wine is then drawn off into huge casks, and about eight per cent. of Brandy (made from the common *consumo* wine of Portugal) is added to check further fermentation. The process has so far only taken about two days; a Bordeaux must is allowed to ferment for anything from two to three weeks; so that it is evident that there must be a tremendous quantity of unfermented sugar in the Portuguese must when this Brandy is added. There then begins a battle between the sugar and the Brandy, the sugar trying to turn itself into alcohol and the Brandy trying to prevent it; and this struggle may last a hundred years. The wine in its arrested development is transported down the river to the lodges of Oporto; it is carried in goatskin sacks to the country adegas, whither the shippers send up their wooden casks, and the Brandy that is used to check the fermentation. Dr. Thudichum saw many German casks, and suspected that much of the Brandy used was German spirit. But this need horrify nobody. Quite recently it was my privilege to drink a Sandeman 1897, a vintage rather eclipsed by its predecessor of 1896, if that be not a bull. But the '97 was as good a Port as I ever drank, and I sang its praises in the pages of *Wine and Food*, the quarterly of the society of that name. This brought me a letter from Mr. R. E. McEwen, in which he told me that the late Colonel John Glass Sandeman, who had been his neighbour for years, had told him that the 1897 Sandeman had been made with Scotch Whiskey instead of Brandy, as there had been a Brandy famine that year, owing to the large amount of their stocks used in making the 1896.

It is now time to recall that Port is a wine of two characters, according to the treatment it receives. A Vintage Port is shipped in such years as warrant the shippers in believing that the wine of that year is so excellent and self-contained that it can be put on the market without any blending or toning up. [CAMPBELL: *A trifle misleading to say 'without any blending,' as a Vintage Port is a blend of wines of one selected year; thus a* 1927 *Vintage Port will be a blend of wines of* 1927 *only.*] It accordingly spends its two years in cask, is bottled and laid down, and waits for Time to bring it to perfection in anything from twelve (a most unusually short period) to ninety (a most unusually long period) years. The wine lies still in the cellar, the unconverted sugar struggling against the Brandy and slowly but surely converting itself into alcohol (I presume that the gas so manufactured seeps out through the cork, for no combination of cork and sealing-wax is absolutely airtight, and if it were the struggle of the sugar would be in vain). The result of this slow chemical change is to be seen in the gradual deposit upon the side of the bottle of what is called a crust; this has its origin in the skins of the grapes, which during the first fermentation partly dissolve, giving their rich colouring matter to the must, but also making with the sweet juice a mucilage which the subsequent chemical process develops into something not unlike the original grape-skin. A Claret will throw a crust also, as will a Burgundy or any other red wine; but the Port crust comes from a stronger, sweeter grape, and is more characteristic of the wine. The squire who beat his parlourmaid because she broke the crust of a bottle of venerable Port, was angry with her for a breach of ritual; but the true foundation of the ritual, and his anger, was the importance of not shaking wine, and thus re-mingling with it all the impurities that it has carefully shed during its rest. As a matter of fact, I think it is usually easier to succeed in the decanting of a bottle of Port than in the decanting of a bottle of Claret. The deposit thrown by Claret is in the form of a fine dust, and it takes very expert judgment and skilful hands to prevent a trace of that passing from the bottle to the decanter. But the crust thrown by Port is probably sticky enough to hold the fine dust deposited by that wine; and all you need worry about is the risk of the crust slipping.

So far Vintage Port. Obviously it must be a costly wine, not only because the very best grapes of the very best years go to its making, but also because it takes so long to reach the table: anything from twelve to ninety years. Even in this year of grace 1940 I have not yet seen any of the 1927's. The 1922's are showing prettily, and the 1912's magnificently. But there are three wines thirteen, eighteen and twenty-eight years old, and the money paid for them locked up all that time. Add the fact that Port is a wine very heavy in alcohol and it will at once be seen why good Vintage Port can never be a cheap wine.

But not all the wine made in even the best of vintage years goes direct to the consumer. The shippers set aside a considerable proportion for the purposes of blending; and the blended wine usually is what is called Wood Port, that is to say, Port matured in cask and not in bottle. Wine in cask is much more exposed to the action of the atmosphere and suffers a good deal of evaporation. The colour is affected; you can usually tell that a Vintage Port has been late-bottled by the lightness of the colour. Wood Ports are known as Ruby and Tawny, according to the colour they have attained. I have never been able to ascertain whether there is any other difference between a Ruby and a Tawny Port. Now, it is important that a Port shipper shall be in a position to let his customers have a more or less unvarying article of commerce, which his Vintage Ports from the nature of things could never be. He will accordingly make out a series of patterns and number them or name them, and he will say: 'These are the Wood Ports that I can undertake to supply year after year to practically any extent.' Each pattern will be the result of a set formula; so many measures of ordinary, or second-rank, wine, so many measures of a wine of a former vintage that was very rich in sugar, so many measures of another that was of a particularly good colour, and so on. The jeropiga plays its part, of course. The resulting patterns will vary in quality and price, according to the nature and value of the component parts. But your club need never be afraid of running short of its standard Shilling or One-and-Sixpenny; it can restock month by month, certain that each succeeding parcel will be the same as the last. And Wood Port need not be treated with the awe appropriate to Vintage Port; it will probably have been bottled just before delivery, and will not have had time to throw a deposit. The householder should buy his Wood Port in small quantities at a time, drinking it as it arrives, and not allowing it to lie in bottle for any length of time. Wood Ports are inclined to develop a sort of bottle-sickness if kept for more than a month or so; and they will then have to wait until they have acquired a certain bottle-age before they are fit to drink.

It will be noticed that Vintage Port and Wood Port alike depend upon added Brandy for their particular nature and quality; whether Sherry was earlier or later in calling in the alliance of distilled wine I cannot say. I suspect it came later at Jerez; I doubt that the sherrisack of Shakespeare's day was a Brandied wine. In those days 'wine' meant Bordeaux wine in England. Gascony and Guienne had for long been part of the dominions of the King of England; and it was natural that the Englishman's glass should be brought from the country that was not only the nearest vineyard but was also under the same crown. The thrill of drinking Sherry came partly from the fact that it was Prize—*prise*, or captured from the enemy. But the discovery that there were wines of a character quite different to those of Bordeaux must have caused a certain inquisitiveness about the possibility of still further variety of flavour and quality. And meantime the history of England had taken a new turn. When at the end of the fifteenth century explorers from Bristol discovered Newfoundland, England was still in allegiance to the Church of Rome, which at that time was passing through not the most glorious period of her history. One suspects that there was a great deal of taking tithes of mint and anise and cummin side by side with the scandals that had already come under the lash of Reuchlin and were almost immediately to cause the defection of Martin Luther and his friends. Erasmus might have saved the situation; but he was comfortable and was not prepared to risk the loss of his comfort. So the schism came; the King of England denounced it, and earned from the Pope the title of Defender of the Faith, which his successors (some of them with greater fitness) still bear; but the King was amorous; the Pope was difficult, and to enable himself to enjoy a new wife for a little while, Henry abolished the rule of Rome in England. The first advantage taken of the new dispensation was to abandon the observance of

Fridays[1] and fast days, then more numerous than in recent years, when fast nights appear to have superseded them. This was a heavy blow to the recently created trade in codfish, whereby the settlers of Newfoundland bought all their manufactured goods from England and paid for them with fish. The English market being thus lost, the colonists looked around for a new market, of firm Papists, preferably; and they found precisely their affair in Portugal, where the inhabitants were not only unyielding in their fidelity to Rome, but also had the further advantage of being subject to a national hallucination that codfish was a delicious article of diet. But alas! all they had to offer in exchange for the fish was their wine, then an unfortified commodity, rather like an inferior Claret. This the Newfoundlanders were forced to accept. Unfortunately, the southern climate had not made for diligence; the Portugal wines were often shipped before their fermentation had worked itself out; and an alarming proportion of their casks reached the other side of the Atlantic with their sides stove out by the action of the gases generated during the voyage. Some genius had the idea of checking the fermentation by adding Brandy; and the conception of modern vintage Port had been born. But only the shipments abroad, and chiefly those for the other side of the Atlantic, were so treated at first. In the meantime history rolled on; Charles II married Catherine of Braganza, thereby laying the foundation of the Indian Empire; he really loved the poor little Portuguese princess, and was willing to do everything for her except remain faithful to her; he gave her her own Court, her own chaplain, her own companions; and when she expressed a longing for the wines of her own country he saw that she was duly supplied. To drink a glass of Portugal wine at the Queen's Court became a fashionable thing to do. Portugal wine became chic; but it remained caviar to the general. Then came the invasion by William of Orange; in the war with France which resulted economic warfare was first used on a large scale and all French wines were forbidden entry to England. In search of a substitute, people said: 'Why

[1] This is not a completely accurate statement; Henry himself was in some ways more Catholic than the Pope, and Elizabeth at one period ordained three fish days a week!

not import wine from Portugal?' And they began to do so. Most of the wine so imported had begun its life in the Gironde, and had suffered a sea-change by being imported into Oporto and exported thence to England. But a certain amount of ordinary native wine was sent over also. People who drank it noticed that it tasted different to the Portugal wine that used occasionally to reach them from friends in Newfoundland; it was neither so sweet nor so strong. Some Scotsmen suddenly appreciated the commercial value of this; they set out for Oporto, and commenced preparing wine for the English market upon the lines on which it was already being prepared for export to Newfoundland. And that is the origin of the modern Port wine trade. You will still find an occasional house where you will be offered a glass of Newfoundland Port, and very good it will probably be. But I fancy that it is merely a term of description nowadays, like East Indian Sherry, which in modern times can rarely have made its pilgrimage to India. Probably in the days when Newfoundland Port first earned its popularity in England there gradually set in a channel of trade through the hands of certain merchants, who have kept up the tradition. But Port, whether it came here direct from Oporto or via the other hemisphere, could never have attained its unshakeable popularity if it had not a sort of natural affinity with the English climate and the English character. It is a wine that lends itself to ritual, a thing always held in high honour in a country which has banished it from its church, just as the Americans, when Democracy had banished all titles, took to being addressed as Colonel, or Major, or Senator, or Judge, or even in desperate circumstances as Congressman. Natural instincts cannot be successively stifled. And Port has much more than its ritual to recommend it. It is hard to find words of praise for the east wind; but it does make a glass of Vintage Port more grateful and comforting. And when in October the nip is coming in the evening air, and the logs are once more crackling on the hearth, what pleasanter occupation could there be than the slow and critical consumption of a bottle of old Port? With bed the next halting-place, no work to be done, nothing to distract the mind from kindly memories of friends that are gone except the good-natured gossip about those who still are with us; with good apples and nuts to give the teeth something to bite on, and supply a proper contrast of flavour, unless this be sought in a good plain seed-cake or similar plain cake—this surely is the English autumn at its best.

But Port must be allowed its vagaries. It has never forgotten that its first popularity in England was due to that Champagne-drinking monarch, King Louis XIV; and it will not brook the society of the sparkling wine without exacting a severe retribution from the drinker's head. With Claret it had no quarrel; Burgundy it can tolerate; but if asked its own views as to the best prelude to itself I am sure it would reply: 'Beer; honest English draught beer.' And it prefers to be treated with the deference that is prepared to refrain from serious employment after the banquet; for which reason it is always better after dinner than after lunch, when vintage Port is inclined to produce a drowsiness inimical to vigorous application afterwards. Those who like Port in the middle of the day would be better advised to stick to Wood Port, which seems for some reason to be more digestible. Probably the access of air to the wine in cask has robbed it of much of its heaviness. But in the same way I should counsel Vintage Port for evening drinking. The resulting somnolence is then a blessing rather than a disadvantage; its majesty is more appropriate to the meal that is meant to soothe and comfort rather than to strengthen and energize.

My judgment on Port was destroyed at the outset of my vinous career by the hospitality of my Colonel in the Royal Dublin Fusiliers, the late Jim Meldon, the handsomest, most generous and in every way the most typical Irishman I ever knew, God rest him. It was he and the late Ernest Page, K.C. (who might well have stood for England as the Colonel stood for Ireland), who were my sponsors for the Windham Club when I joined it in 1922; and incidentally it was at Ernest Page's hospitable table that I first met Newfoundland Port. But it was Jim Meldon who 'dined me into the club.' At that time the Cockburn '96 was 'on Tap' at 2s. a glass (I mention the price for the sake of comparison of those days with these); and he insisted that we should take double glasses after dinner. I can still see the twinkle in his eye as he set down the empty glass with the remark: 'You know, it slithers down!' It certainly did slither down; and never have I tasted a Port so much to my liking, although the Cockburn 1912 has occasionally come near it. But my complaint against my host, if complaint it can be called, is that he did not warn me that I was drinking probably the finest Port I should ever taste. In due course I met others, probably very fine wines also; but they were not Cockburn '96, and therefore to my inexpert palate they appeared unworthy of praise. Alas! to-day the wine has passed its grand climacteric; let nobody who has only met it in the last few years think he has tasted it. Imagine a mere robust and yet a more elegant version of the Cockburn '12, and you will get some idea of what it was at the age of twenty-six. We had some Taylor of the same year, and the connoisseurs usually accorded it a higher place. But I never affectioned it in the same way. Both wines now appear at the Bench table of the Inner Temple from time to time, and again those whose opinions I value set the Taylor above the Cockburn; but still I disagree, and Cockburn remains my favourite shipper. Sandeman I find on the sweet side, but for those who do not consider that a fault it is always a safe choice. The Warre 1900 has given me several pleasant moments; and there are more than a couple of shippings of Dow's that I remember with gratitude. Croft has also earned its benediction; I never cared very much for Offley, a shipper that commanded the firm allegiance of that formidable judge of wine and men, the late Marston Buzzard, K.C. At the moment (or at any other moment) I have nothing unfavourable to say about Fonseca, whose 1922 is worthy of repeated and attentive examination. Burmeister I never liked; on the other hand, I have never drunk an inferior Graham.

With the disappearance of our supplies of French wines I suppose that it will have occurred to many people that there might be unfortified wines from Portugal which might 'keep the bed warm' for our Clarets and Burgundies until the day when order is restored in Europe. But it is not as simple as all that. You cannot build up a wine trade in a day. If England wanted Portuguese Clarets, it would be necessary to plant new vineyards, producing ten times what is produced to-day; and these would not attain excellence for several years, nor would they then be giving us a wine as good as the wines of France. And suddenly the French wines would return; and all the Portuguese expenditure of labour and capital would have been wasted. So I do not propose to say anything about such wines, beyond saying that about five years ago I had a few bottles of a sort of Claret from Portugal of the vintage 1906, which made a not undistinguished prelude to a Château La Lagune of the same year; and recently I drank at the hospitable table of Messieurs Reid, Pye and Campbell a white wine, old in bottle but of unknown vintage year, that might have played substitute for a fairly good Chablis. It was very clean and dry, with a good, fruity flavour; and it had a very beautiful colour, not the least of a wine's attractions. So if you have any friends in Portugal——!

THE WINES OF ITALY

IT is probably not the most appropriate moment to choose for saying a word of praise of Mussolini, yet I trust that the baseness of his conduct in recent years will not altogether obscure the remarkable and extensive reconstruction of the Italian nation for which he is entitled to a great deal of the credit. I had, indeed, hoped that this material achievement was only the outward sign of a moral regeneration; and that the people of Italy were daily advancing to higher standards of citizenship as well as towards better methods of commercial enterprise. And in particular I was awaiting the moment when he would realize the potential wealth lying dormant in his vineyards, and start a reform of the wine trade. For Nature has endowed Italy much more richly than France with the gift of vines; the traveller cannot remain unaware of their omnipresence; and probably because of their prolificity they are treated as though unworthy of more than the most casual attention. I have from time to time drunk the most lovely wines from Italy, wines quite worthy to play substitute for the lesser growths of Bordeaux or Burgundy. [CAMPBELL: *I think Maurice ought, perhaps, to have said* 'in *Italy.' We never get 'most lovely' Italian wines over here. Further on, Maurice modifies his enthusiasm by saying they may be drunk* on their own doorstep *with pleasure. The Italian wine growers do not, as yet, seem to have laid themselves out seriously to make wines fit for export to Northern climes and consequently such wines as they do send us are inclined to be* agro-dolce *and crude and unlikely to age well. A big reward might await a greater and more sustained effort. Some years ago a wine merchant in Scotland wrote and asked me if I would 'vet' an article on wine which one of the most prominent doctors of his city had promised to write as a preface to his new price-list. I consented and a few weeks later received a proof of the literary foreword which gave a general approbatory survey of wine with an occasional historical or slightly erudite allusion. My friend sent me a covering letter in which he ventured to hope I would agree that the article was excellent, but, he asked, 'has not the doctor made a mistake in mentioning a name Falernian? I have none of it on my list,' he wrote, 'and have never heard of such a wine: what do you think he means?' My reply was very tolerant and begged him to leave the intruder in. I told him of green vineyards on the slopes of Vesuvius and the passion an old friend of my boyhood called Horace had for their wine. The wine-list must have been a colossal success as, a few years later, my wine merchant friend sold his business at a very tall figure.*] But usually when one calls for a flask of Chianti or Orvieto the contents prove to be a wine ill-made, carelessly bottled, and badly stored; and the only way in which the French wines have been successfully imitated is the price. The vines are allowed to grow wild and untrained; they soar to ridiculous heights, regardless of the fact that no good wine can be made from vines which constant and careful pruning and training have not kept close to the ground; they flourish in fields where other crops are allowed to take toll of the soil, and in some strange way to affect the flavour of the grape. When gathered, they are fermented heedlessly and without care; they are kept with an equal disregard of the rules that have made the perfection of French viniculture. They are bottled haphazard; in Tuscany, the primitive *fiasco* was, until the other day, only stoppered by a drop of olive oil and a wad of tow to keep it in place. Heaven alone knows what strange guests are sharing the flask with the young wine. Nevertheless, so beneficent and powerful is Nature, these wines made in this heedless fashion can often be drunk on their own doorstep with genuine pleasure. It is only when they come to the adventure of a long journey and a foreign storage that they reveal their imperfections and fade away into tasteless and vapid liquid.

It is the grape that gives flavour to the wine; and if the grapes have been ill-grown their own flavour will have been impaired. If they are allowed to proceed to the vat without due care having been taken to ripen them properly, and subsequently to detach all imperfect berries before setting them to ferment, the chemical transformation that then results can supply no merit already absent, and will not remedy their imperfection. If they are launched on their career thus imperfect, the period of maturation in cask will be more than ordinarily perilous by reason of the many risks of fungus and 'vinegration,' if I may coin a word, that all wines must fear, and the badly made more than any. And, bottling being a difficult and delicate art, only to be attempted in favourable climatic conditions, it stands to reason that the dangers run by the best wine will be exaggerated for the ill-developed wine. The greatest genius can hardly assert his native talent if he has the ill luck to pass from the hands of a careless nurse through those of an ignorant schoolmaster to a university of no distinction.

It is for this reason that our experiments with Italian wines in the neighbourhood of Soho have so often proved disappointing. The one advantage towards successful salesmanship that these wines possess is the attractive form of their bottle. The straw-covered flask in which your Chianti reaches you is a poetically shaped and christened container; it makes a special appeal to feminine tastes; and it can be put to various ornamental uses about the house. Besides, those who like that sort of thing are not often well equipped to criticize and appraise the contents; and if the *pasti* and the *zabaglione* are good, the wine will serve to wash them down. But the wistfulness with which the Cook's Tourist still vainly seeks that delicious little wine he drank last year at San Gimignano is the cause of a certain atmosphere of melancholy in some of the best Italian restaurants in Soho, the proprietors of which are not to be blamed if they have failed to repair the results of the carelessness of the owners of the vineyards whence come their Chiantis and Barolos.

Italian wines are very often classified, not by the district names, but by the name of the vine from which they have been made. A vine, such as the Nebbiolo or the Monferrina, may be grown in several different parts of Italy; and the wine made from it in one province may be entirely different from that made from the same vine in another province. But both wines will be sold under the name of the vine, the name of the province being added as a qualifying description. It is much as if we sold our beer as 'Bitter—Truman's' in contrast to 'Bitter—Tetley's.' I do not suppose either brewer would consider it a reflection on his product to be told that it was quite a different thing from the other, although probably both beers are made from the same ingredients in very much the same way. But the water, for example, is different; and in the same way the difference of soil between one part of Italy and another can make wine fermented from the same species of grape taste quite different in each district. For this reason it would be ridiculous to write of Italian wines in the same way as of the wines of France or Germany, Spain or Portugal.

Let me start plump, therefore, by saying that my own experience of Italian wines (a very slight one) puts the wines of Verona far and away above the rest. In 1926 when I visited Venice my friend George Washington Cass Papazian took me to a restaurant called Vapore's, where the only other guests that day were Diaghileff and Lifar, who were lunching together with much laughter like a pair of schoolboys, which perhaps is not a bad appreciation of the one and the other. Nice schoolboys, of course. We had an admirable lunch; and we drank a red wine which the proprietor told us came from Verona, and was a mere *ordinaire*. It was the first time that I drank it; but it was not the last. It had all the Claret quality, coupled with a warmth of its own that I can still recapture in my mind. It was completely limpid; it had attained sufficient bottle-age to be ruby red without a hint of blue; and the taste of the wine had developed equally, so that none of the slightly vinegary flavour of a young wine had survived. If I am right in my recollection, its price was only the equivalent of about 1s. 6d. a bottle. It was in a wine-producing country, but it was also in a restaurant; so that I think I can truly call it cheap. From the time I first met it I always sought a Verona wine in Northern Italy; and I cannot remember ever being disappointed.

In London my principal acquaintance with Italian wines has been formed at the Isola Bella Restaurant, where I have been a constant and frequent visitor for twenty-one years. Commendatore Micotti has been interned for the duration of the war, and I suppose that no criticism will be aimed at the Home Secretary for his action towards one whom Mussolini honoured so recently; but I am sure in my own mind that it is possible for an Italian to be a good Italian and completely friendly to this country also, and I hope the day is not too far distant when I shall be welcomed again to 15 Frith Street by the proprietor himself. My beloved Giuseppe Bonini, the head waiter with the *kaiserlich* moustache, who had been my good friend for twenty out of the twenty-one years, had gone back to Italy in poor health a few months before war was declared between England and Germany; otherwise I suppose he too would be behind the barbed wire just at present, and I am very glad that he is not. It was under his attentive care that I first met my publisher and planned a previous effort; my feelings towards the house remain kindly.

But I was talking about the wines of Verona; and I dragged in the Isola Bella because a few years ago Mr. Micotti suddenly 'let me in on' a private supply of Val Pollicella, one of the principal Veronese wines which was, I think, of the vintage of 1927, which had been so dreadful all over Europe, except in Portugal. Whatever the vintage was, that wine graced my table on all occasions subsequent to the visit on which I met it; and I never offered it to a guest who did not praise it. Indeed, with the generosity which so many people have noticed as my chief characteristic, I asked Mr. Micotti to give a similar privilege to my friends; with the result that the supply ran out with alarming rapidity and recent conditions made reinforcement impossible. But that wine had much of the quality that had recommended the wine at Vapore's; and it had more breed and distinction about it. I have never met it elsewhere in London.

But I met many other Italian wines there, which put me on the track of their cousins when I reached Italy. For some reason, Italian wine means Chianti to most Englishmen (or perhaps to most Englishwomen, in which case the *fiasco* is the probable explanation). Now, I was as susceptible as anybody else to the idea that when eating Italian food an Italian wine is indicated; so I began with Chianti, and went on to Barbera and Barolo and Orvieto (not all at the one meal, I hope you understand). And the Chiantis were wonderful—in their variety. The outstanding feature was the way that they were always of years that had been failures in France. In recent years 1931 has been a most frequent label on a Chianti bottle. Of course, these years may have been successful in Italy; and it is only right to remember that the vintages of Italy are more constant than those of more northerly countries. But for a period I had some hopes of the Chianti made and bottled by the Baron Ricasoli, which I had first met in Florence, where the proprietor of the restaurant had given me a glowing account of the praiseworthy efforts made by the Baron to improve the wine industry. I think that in Florence I drank his wine from the usual flask; but when I found it in Soho it had taken to the Claret bottle, in the neck of which its cork was secured by a cross-strapping of tricolour tape before the seal had been put on. The label was also rather splendidiferous. The wine was good enough for me to buy some for home consumption; but I regret to say it did not turn out a successful adventure. In a short time, what I can only describe as a dirty taste started to assert itself. It may perhaps have been due to some fault in my cellar; but if so the fault did not affect any other wine. And I noticed that the Chianti Ricasoli vanished from the Isola Bella wine-list; so that I think there must have been some falling off due to the voyage.

It is my own belief that nothing in the world could succeed in making Chianti into anything more than a passable wine; whereas I hope some day to taste a Verona that has had the benefit of Baron Ricasoli's care and attention, for I believe that the Baron de Brane could have turned the Val Pollicella that I knew into a wine of real distinction in a dozen years. I would also like to drink a specimen of a nursed and cosseted edition of a Grignolino that I once met in Naples, which was about as jolly a little red wine as I have ever drunk. It kept winking at you over the straw covering of the flask and persuading you that it was time to fill your glass again. It had something that I should call a sparkle if that word had not a special meaning with regard to wine, and this wine was absolutely free from gas. But there was something that gave a jaunty set to your hat as you drank it; and it gave you the impression of being saucy without being impudent. If that Anglo-Irish phrase, 'a broth of a boy,' means what the average Englishman thinks it does, then this little Grignolino was the broth of a boy. I have never drunk Grignolino in London, and I have a suspicion that it does not travel successfully.

Barolo has always seemed to me to fall between two stools it has not the finesse of a Claret type, nor has it the robust quality of the Burgundian. The latter is better understudied by Barbera, which, however, I have always found too coarse for my taste. And all the Italian wines that I have met suffer from the imminent possibility of that dirty taste of which I have already spoken announcing its arrival in the most innocent-looking bottle.

Orvieto and Suave are about the best of the Italian white wines; and I do not care much for white wines. I know that there are many other well-known names, which are familiar to every traveller; but in this book I am more concerned with the wines one drinks at home; and I do not think that one often meets any others than those I have already named. 'Lacrima Christi' is a name known to most people; applied to any wine, it seems blasphemous to me (and I dislike the name 'Liebfraumilch' also) but stupid as well as blasphemous when applied to the ridiculous liquid marketed under that label. There are pleasant wines made on some of the islands near Naples; there is Vesuvian also, a hellish brew tasting of Harrogate waters. But there is one more wine, worthy of more serious attention, and linked historically with one of the noblest names in the story of this country. I refer to Marsala.

Marsala is a fortified wine, made much as Sherry is, but not plastered. It is, or was until recently, very cheap; it is, or was until recently, very strong; and Horatio Nelson, in the intervals of paying court to Lady Hamilton, ordered that the British Navy should be kept supplied with a liquid that looked like and was nearly as strong as rum and cost a great deal less. When Marsala is good to begin with, and has been kept for the requisite number of years, it can be a truly noble wine. When I joined the Windham, they had an Old Particular of Woodhouse's that only cost 9d. a glass, and was indistinguishable from a very fine old Madeira. I am bound to say, I have often since wondered whether I did not enjoy the benefit of somebody's mistake; for I have vainly sought for another sample of such Marsalian excellence, and it *has* occurred to me that a bin in the cellar might have been mislabelled. If the bottles happened to be unlabelled (as they probably would have been) I see no reason why I might not have been drinking Madeira! But although I never repeated this paragon, I did find very worthy old Marsala, and I should never refuse it cellarage.

MADEIRA

ALTHOUGH Madeira has gladdened the hearts and palates of mankind for centuries, the only book that I know which deals with it is André Simon's volume in Constable's Wine Library. My principal criticism of my beloved perceptor is that he is inclined to assume too large an acquaintance with the subject on the part of his readers; and on the occasion of the preparation of that book he gallantly left half the space to Miss Elizabeth Craig, who gave us many useful hints upon the proper use of the wine in the kitchen. But I wanted to know more about the wine itself. And Doctor Thudichum had dismissed it with a wave of his hand; the older books are still more unsatisfactory. Mr. Sheen, writing about eighty years ago, lumps Madeira with Canary, and throws in a bit about South American wines in the same chapter. Cyrus Redding a little earlier had done somewhat better by the wine, but even his thoroughgoing method had not fully developed the subject. And Mr. Morton Shand, my constant standby, gives a mere four pages of his book to a wine that deserves to rank in relation to Port and Sherry as Trinity College, Dublin, stands with Oxford and Cambridge—junior, it may be, but none the less distinguished. Personally, I hope some day to see a volume produced by Sir Stephen Gaselee upon the wine that he understands better than any living man, and the great qualities of which he has so generously revealed to his friends. [CAMPBELL: *We all talk of Stephen Gaselee when the Madeira comes round, and yet it only occupied a corner of his massive and kindly brain. He was of those who could ill be spared.*] It is with diffidence, therefore, that I venture to put together a few paragraphs about a wine that I esteem and cherish, but hardly know; for the mere fact that I have been privileged to taste some of the island's finest vintages is about as good a qualification for writing on the subject as would be a tourist's visit to Sèvres in respect of the famous porcelain.

Madeira is a fortified wine, and is, as regards the tradition of its making, a descendant of Port. The island of Madeira was discovered and colonized in the first quarter of the fifteenth century. It was then a heavily wooded and almost uninhabitable terrain; for some reason the new settlers, whether through laziness or through inspiration, did not adopt the obvious plan of clearing the timber and selling it; they set the woods on fire, and for seven years the island was a continous blaze, until at last the fire burned itself out, leaving a soil free from all noxious forms of life and enriched by the ashes of the conflagration. Why the Portuguese should have gone to Crete for the vines with which to plant it is hard to explain; but this is the reason that Madeira is a derivative from the old Malvoisie, or Malvasia vine. This is the vine that gave us Malmsey; but the Duke of Clarence's vat had not come from Madeira; it had probably come from Eastern Europe.

There are three main classifications of the wines of Madeira; there is the very rich Malmsey, there is the rather sweet and rich Boal, or Bual, and there is the more austere and majestic Sercial, which is the wine with greatest lasting powers. All share a dominant flavour which is, so to speak, national; each has its own peculiar distinctive taste. Some people choose to find in what I call the national flavour a taste of pineapple. I have not visited Madeira, and I am unable to ascertain whether the pineapple is a local product; if it be, its penetrating aroma and tang would be well calculated to permeate everything in the neighbourhood. I confess, however, that I have never thought of pineapples when drinking Madeira.

Madeira became a regular inhabitant of English cellars about the middle of the eighteenth century. It provided a variant of Port, and satisfied the demand for a fortified wine which had followed upon the cultivation of the Port ritual. A number of British merchants became associated with the island, and the development of the tourist habit brought Madeira so much upon our travel map that we have to think twice before realizing that it is not a British possession.

The merit of Madeira was always great; but, like most

other wines, its golden age was before the great plagues of oidium in 1852 and phylloxera in 1873, and the new vines do not seem to be able to produce the lasting wines that used to be associated with the old. [CAMPBELL: *To my taste, Madeira must be old to be enjoyable. When it is really old and really fine it is matchless.*] Seven or eight years ago the Saintsbury Club were fortunate enough to become the possessors of a parcel of a Madeira of the 1792 vintage, which had been purchased for Napoleon's use in 1815. He never took delivery of it, and the English Consul at Funchal, Mr. Veitch, was left unpaid. In due course he took re-delivery of the cask, and sold it in 1822, after the Emperor had died, to Mr. Charles Blandy, who passed it on to his son John. The latter bottled it in 1840, in which year was born Doctor Michael Grabham, whose death at just short of the century occurred quite recently. In due course the Doctor became a suitor for the hand of Miss Blandy, the daughter of John; and when he married her he received the 1792 wine as part of the marriage arrangements. When I first tasted this wine it must have been about one hundred and forty years old; it appeared to have gained rather than to have lost strength; and it certainly lacked no quality that one associated with a noble wine. By the grace of Sir Stephen Gaselee we were sometimes able to compare it with a younger wine, believed to be an 1812, but certainly earlier than 1850, from the estate of Donna Isabel Esmeralda da Camara; the latter was a Sercial; I think Napoleon's wine must have been, also. But even the réclame of its imperial ownership did not promote the 1792 over the head of the younger wine, which always seemed to me the most delicate and beautiful specimen of Madeira that it has been my privilege to drink. And Sir Stephen is determined that the memory of Professor Saintsbury shall never be drunk in unworthy Madeira; from time to time there reaches our cellars a delivery of wine of slightly more modern vintages, but all pre-oidium and venerable, which he has managed to find during his visits to the island.

For some reason those who are privileged to know Madeira in this magnificent guise are usually doomed to bitter disappointment when they go on the market as purchasers themselves. I have occasionally asked to be supplied with a Madeira really worth while; but, although I expect to be believed when I say that such requests were addressed only to first-class wine-merchants, the resulting wine was invariably unsatisfactory. Mind you, I was not proposing to pay a pound a bottle; but I see no reason why the best Madeira should cost more than the best Port, and I am speaking of a time when I could have bought a good vintage Port at about 150s. a dozen. I therefore think I should have been able to buy a fairly good Madeira for about 90s. But all I tried had a curious back-of-the-throat flavour that was sour and nasty. About eight years ago the Madeira Wine Association took the matter in hand and put on the market at a very reasonable price a pleasant preprandial wine which they called 'Gloria Mundi'; I think it cost about 7s. a bottle, and it was certainly excellent value for the money. I do not think I tasted their dessert wine which they called 'Duke of Clarence Malmsey,' and claimed for it the useful quality that it was non-gouty. I write of them in the past tense, for they have vanished from the Telephone Directory, and I do not know what has become of their marks; I trust that they are still to be found elsewhere.

A word or two about the wine itself and its manner of making. The soil of the island is volcanic, and thence comes the affinity which the humbler Marsala can boast. It is also waterless; and a most elaborate system of irrigation has had to be set up to keep the estates in water. This is delivered through canals at stated times and with the inexorable regularity we generally associate with a dictator; the owner who is not ready for his supply at the appointed hour must wait until his turn comes round again. This is important to remember when measuring the worth of one estate against another; a vineyard under the care of those who are punctual about receiving and restoring their water supply will naturally produce a more regular and better crop than one where the supply is missed one week out of four. By the way, this practice has nothing to do with 'Rainwater Madeira'; the latter is an American term, applied to a wine imported by a gentleman named Habisham, and matured and refined by him in a secret manner of his own so as to be lighter in body and colour, while unimpaired in bouquet or flavour.

This latter qualification prevents the obvious suggestion that the addition of a quantity of rainwater to Madeira would naturally cause it to be lighter in body and colour; but I fancy that this was the thought in the minds of those who first invented the name.

I have said that Madeira is made in the same way as Port. I have seen it asserted that the wine is not 'plastered'; but Mr. Redding says that 'when they fermented their wine the growers are described as bruising and baking a certain stone called *jess*, of which nine or ten pounds were thrown into each pipe.' This sounds to me like the addition of *yeso* to the Sherry must, which is the operation known as plastering, and which, let me repeat, is merely an antacid precaution. Indeed, its existence might support the claim, just recorded, that Madeira is a non-gouty wine. If I were asked to express an opinion based upon the taste alone I should say that the wine *was* plastered; I think I get the dryness of a Sherry always present, even although it may be masked by the overlying sweetness of a Malmsey or a Bual. With this exception the vintage proceeds as in Portugal until the wine reaches the storehouses, where it undergoes a treatment that is unique (except, perhaps, at Château Haut Bailly, during Monsieur Malvezin's ownership, when he practised pasteurization of his wine). The process is known as 'stoving,' and consists in storing the wine in a cellar or storehouse in which a temperature of anything from 80 to 120 degrees Fahrenheit is maintained. The higher the temperature, the shorter the treatment; it is supposed to age the wine, and the practice has produced some strange experiments. Some owners have been known to bottle their wine and bury it in a trench filled with fermenting horse-dung; but such practices are not encouraged by the better-class of growers. It is traditional that the wine is improved by a voyage, like East Indian Sherry or Newfoundland Port; but it is said that a steady rocking by mechanical means will mature the wine just as effectively as the motion of a ship. After all, the scientific principle of the voyage is that the motion of the ship will roll the cask, and cause the suspended particles of sugar to dissolve themselves into their component gas and alcohol. There is no reason why this result should not be achieved by a machine.

Madeira is a fortified wine; about 10 per cent. of Brandy is added to the must, and sometimes another dose is added before exportation. But it is not as strongly Brandied as Port; and on the other hand, it gains rather than loses in alcoholic strength as it grows older. As the alcohol, other than the added Brandy, must come from suspended sugar, it follows that a young Madeira is a most deceptive wine, and is wasted if drunk in that condition. I have never made up my mind as to the best setting for Madeira. Not as an *apéritif*, I think; it never really defeats Sherry on this *terrain*. With the soup it is admirable; but not too fine a wine should be used for such a purpose. On the whole I am inclined to reserve it for post-prandial drinking; its richness at its best enables it to frown down the pretensions of any but the greatest Ports, and an old dessert Sherry is its only real competitor. Now that we are temporarily cut off from all countries exporting natural wines I have no doubt that the desire for variety will revive the interest in Madeira, which must always be counted as of the blood and the worthy table companion of every form of natural or fortified wine.

A MIXED BAG

WE have been through all the vineyards of Europe from which come our usual beverage and dessert wines; there remain one or two other products of the vine that call for notice. First, let me mention the wines of Switzerland, if only because I associate with a Christmas holiday in Caux my first acquaintance with great Claret (a Mouton, in magnums; probably the 1899, but in my ignorance I did not ascertain the date, and to work out a vintage from the memory of its bouquet, appearance and taste is a task to be compared with those of Hercules. It was old enough to have a touch of the onion-peel colour; it was full and majestic in flavour; and it had a suggestion of its promised velvet in its bouquet. So I think it must have been an 1899). We used to use the local white wine for daily drinking, and a Dézaley for Sundays; on notable occasions we fell back on the wines of France. But the Swiss vineyards did not disappoint us; they were more like Burgundy than Bordeaux, and might have understudied the lesser Chablis and Pouillys with grace and distinction. I do not remember any red wines. But the Swiss are evidently good buyers. They stock their cellars for an international clientèle, and a gilt-edged clientèle at that; and good innkeeping causes them to refrain from making the cellar carry more than its due share of the working expenses of the house. I am sure that those magnums of Mouton would have scared us if they had cost £1 a magnum; but the same wine was selling in London at that period at about 144s. a dozen bottles, or considerably more than £1 a magnum. At a restaurant the wine would have cost 35s. a magnum at least. Yet here was the same wine in a jolly little hotel above Caux being offered to us at about half that price. Nor does it rest there. The handsomest couple I ever saw joined in Holy Matrimony spent their honeymoon some half a dozen years ago in a similarly unpretentious Swiss hotel. Finding that they could enjoy a bottle of Château Yquem 1912 at the equivalent of 7s. a bottle they had one every night. I expected my godchild to be a beauty; but the transcendence of that beauty has in my mind been partly attributed to the Yquem, so generously provided at less than a honeymoon price!

I make a wild leap from Switzerland to Czecho-Slovakia to praise the only wine of that country that I tasted. It was a Melnik, and it came from the Lobkowitz vineyards (I thought of Beethoven's uncouth attack on his patron: 'Ass of a Lobkowitz! Dolt of a Lobkowitz!'). The wine was not served with any fanfare of trumpets; I think it pretended to be little more than an *ordinaire;* but it had the most beautifully clear limpidity that I ever remember; it looked against the light like the very best kind of red ink! [CAMPBELL: *Not a compliment to any wine. I must confess I have not yet seen a Czech wine that took my fancy. Hungary could give them a long start and beat them even without the assistance of Tokay. Limpidity is not enough: water can be limpid and yet tasteless (if not chlorinated).*] The taste was far different; it was a perfumed wine, and it caressed the palate. It was of the Claret kind; but it had something of the aromatic quality of a Port, not too assertive, but improved by the suggestion of the Orient. It was not very old when I drank it; about six or seven years. Of its lasting powers I am unable to speak; I have found that wines have great difficulty in attaining great age when I am around. But I have never seen it in England.

The wine most like it in my experience was the Austrian Voslauer, grown and made at the gates of Vienna. Here, if you like, is a jolly little wine, all gay, like the girls in their festival dress, and just as ready to be happy with you, and to cheer you up. The perfume and taste of Voslauer are not the same as those of the Melnik; they (the Melnik) are more patrician and distinguished. But there should be a wine for every occasion; and, just as I should prefer to make Mosel rather than Rhine wine my standard drink, so should I prefer to drink Voslauer on weekdays and keep the Melnik for Sundays. The Voslauer is a Viennese wine; its entire quality is expressed in that.

I had two other experiences with Austrian wines that

amused me. The first was at a little hotel at Urfahr, across the river from Linz; the name was the Achtleitner, in case you are ever able to go there again, for I think it was the best and cheapest hotel at which I have ever stayed. We dined in the garden restaurant, and I asked the waiter to bring me a good wine of the country. He asked me did I prefer red or white, and I plumped for red. 'Then,' he said 'you had better drink Ruster: a half-litre, yes?' 'Oh, a litre,' replied the rash investigator. Now, Ruster is a wine that comes from what had formerly been Hungary; it is a dessert wine, which the waiter ought to have told me; and, in spite of what he indicated, it is what I would call a white wine, although its colour is rather like that of Madeira. But it has nothing red in it and is misused if treated as a beverage wine. However, my companion and I polished off that litre in record time, and I am not sure that we did not follow it with another half; but I know that we slept very well that night, and I awoke on a gloomy, drizzly morning, feeling charitably disposed to all mankind. Few wines have had so remarkable a tribute paid to them.

It was at lunch-time on that following day that I made my second rash experiment. The drizzle had lasted; we were going down the Danube to Vienna; and the charms of the river route demand a fine day to be appreciated. There was so little of the accommodation that was protected from the rain that it was difficult to avoid the attentions of forward neighbours. And there were two gentlemen, father and son, the memory of whom occasionally surges up in my mind when I read of Hitler's atrocities. For they were determined to instruct me; and their method of doing so consisted in borrowing my Baedeker, an English edition, glowering at its pages for a few minutes, and then returning it to me with the assurance that they did not speak or understand English. They then would pour forth a flood of gutturals which apparently had something to do with the Danube, or the surrounding landscape, or some village or schloss that we were passing, after which they would make gestures that were offensively friendly, and smirk in a most self-satisfied way. The smell of the engine-room, which informed the atmosphere of one of the places of refuge which I had sought, although it disgusted me, had no deterrent effect upon them;

and I could think of no other sanctuary. In despair I invaded the saloon, where an early lunch was just about to start; but the place I chose had two vacant seats opposite it, and my faithful guides proceeded to occupy them. I was not hungry, and the table manners of my neighbours were not calculated to improve my appetite; so I thought I would merely drink something, and I asked for the wine-list. And there I found an entry of Constantia, a wine I had never tasted; and without a moment's reflection I ordered a half-bottle. It appeared that there were no half-bottles; but it was too late to retreat now. I ordered a bottle and prepared for the worst.

Now, Constantia is a wine that comes from the Cape [CAMPBELL: *This paragraph is a travesty, a joke. When my father, and Maurice's, were young men, Constantia was a fashionable wine, and I believe it is now coming back to favour again. Still, it came in useful for the story*]; it is a dessert wine, extremely sweet, and had it been genuine I might just as well have ordered a bottle of Banyuls. But all I knew about it was what I had read in Saintsbury's *Notes on a Cellar-Book*, and that I did not remember very clearly; and when the bottle arrived there was nothing about the exterior to put me on my guard. I poured out a glass; it fell like treacle from the bottle, looking like blackberry-juice. I tasted it, for I will never condemn any man or thing untried. I discovered that the appearance was an inadequate prophecy of its sweetness, and the blackberry hue a completely misleading expression of its fruitiness. It was horrible; horrible. I leaned forward, and with a gesture of great courtesy I filled the glasses of the two gentlemen. With profound obeisance I lifted my own glass and pledged them; I then affected to see somebody calling me on deck and I fled. Later I heard sounds of revelry from the saloon; my erstwhile guides made sporadic appearances, wearing one another's hats, and disguised both in drink and in strange clothes of the caterers. One of them got near enough to me to offer me the equivalent of sixpence for the Constantia, but I had no difficulty in persuading him that the wine had been my humble thank-offering for the profound learning with which he had instructed me. And whence it came, or what grape (if any) made it I shall never know.

Hungary has given me several agreeable glasses. It was

at Salzburg that I first came across a Château Paluguay, which had distinct affinities with a good Claret; and the label, I think, made the claim that the wine was in some way related to Château Lafite. The Rothschilds had no occasion to blush for this pretender; I had several bottles of it, and they were all good. Then I got on to Bikaner, a bootier and strong-tasting wine; this did not please me so much, but it had the merit of being reliable, and when I found it at the Hungaria Restaurant in Lower Regent Street there was no doubt about its being the same wine. 'Bull's Blood' was a still fruitier and more powerful wine; I suspect the presence of phosphorus, but I confess that I enjoyed such trials as I have made of it. But Hungary can rest content with the reputation that has been earned for her by the unique wine of Tokay.

Sceptics who read the lives of the saints in order to sneer at the miracles therein recorded should read Mr. Charles Walter Berry's little book on Tokay, where they will find stories quite as remarkable and quite as difficult to believe. He has one story wherein he claims that it actually brought a man back from the dead. The son-in-law of the patient, himself a doctor, came to Mr. Berry for a bottle of his finest Brandy, as his father-in-law could keep no food down. Mr. Berry recommended him to try Tokay instead, and gave him a bottle of the precious Essenz on the express condition that he would give the patient some, no matter what his state. The promise was given; but the sick man was asleep when the doctor returned, so he postponed its administration. During the night he was awakened by the news that the patient had died. He rushed to his bedside and administered the usual tests. There was apparently no doubt about it. And at that moment he caught sight of the bottle of Tokay. Remembering his promise, he forced an egg-spoonful of the wine between the dead lips. After a moment he thought he felt a tiny thread of a pulse; he repeated the dose, and repeated it again; and at the time of the telling of the tale the dead man was a hale and hearty individual. (I have not gone to look up the facts as the book is in the other room, and one cannot always be bound down by the exact text; but I do not think I have mangled the details.)

Tokay is made from the Furmint grape, which is grown

in a volcanic soil, and is peculiarly rich in phosphorus. There is some difference between the various accounts of the way it is made, and I have not seen it made myself; but I think that this is not a very erroneous record of it. The grapes are grown to a very great degree of ripeness, until the skins are bursting with exuberant juice. They are picked one by one, and they are allowed to remain in a container while their own weight causes the juice of the best and ripest grapes to drip into a receiver below. This juice so collected is the *essenz*. Of the grapes not up to this first standard of excellence an ordinary wine is made; but into each vat of this that promises well there is put a ration which may be one, two, three, four or five hods or puttigs of the pulp of the grapes used to make the Essenz, the quality of the wine so made being indicated on the bottle by one to five buttons. A less distinguished wine is made without any added pulp; so that there are really three kinds of Tokay, and it is only of the Essenz that the miracle stories are told. But the others have their merits also, although I myself do not care much for the flavour, and would not cross the road to drink it, except medicinally.

I have never drunk Bulgarian or Russian wine [CAMPBELL: *The Russians tried very hard to get us interested in their Caucasian and Crimean wines during the Second World War. They showed us one or two quite decent light, white wines, but most of the others were poor imitations of other wines, ports, sherries, burgundies, and so on. Their prices, too, were fantastic, and the Food Ministry and Wine and Spirit Association together turned them down uncompromisingly. I feel sure they could produce acceptable wines at acceptable prices if they really wished to*]; but eleven years ago I did a little execution amongst such wines of Rumania as I could find. I had no introductions to anybody in the wine trade, so that I had to rely solely on my own observation and uninstructed local informants; but I gather that the entire wine industry of the country is in the hands of a national company called Capsa, and that most of the wines are blended. I liked particularly a wine of the Claret kind that purported to be an 1899; it resembled a Bordeaux of that year, although not one of the classed growths. I liked this well enough to make it my standby during my visit, so that I have no notes of any other red wines drunk in Bucarest although there were some agreeable white ones.

Greek wines also have escaped me so far; I am told that they are still mixed with such substances as sea-water and tar, which does not sharpen my regret for not having tasted them. But those who are familiar with them say that the flavour grows on the drinker and is much esteemed. I suppose that in due course Saint Simeon Stylites came to think the top of his pillar a most interesting place, as indeed it is to me in speculation. For how did he get his food up? and what about—but perhaps it would be just as well to leave some matters unexamined. [CAMPBELL: *Greek and Cyprus have both improved immensely since economic necessities brought them within our purview. I am not enamoured of them, but Commanderia from Cyprus can be quite attractive as a sweet beverage red wine. It can also be quite the contrary!*]

I once tasted a very agreeable red wine of Burgundy character that came from Chile, and I believe that the wines of South America are by no means to be despised. But this is because the first colonists had a wine-growing and wine-making tradition behind them; the vines were true Pinot and other high-class plants, and the growers took pains with their vineyards, which were planted on a soil suitable for viticulture. About the wines of North America I am unable to speak; before the curse of Prohibition had descended on the land I had never met them, and when America went dry I believe that the wine grapes were replaced in the vineyards by a fruit more fit for eating than for drinking. It will probably be many years before the wine-grape has developed its old degree of excellence, whatever that may be; and the first wines of the old quality are hardly likely to be fit for drinking in my day.

But whatever may be said about the wines of America I am sorry that I cannot praise the wines of the Empire. I am told that South African Hocks are not without merit; and as my informant was that keen connoisseur, the late Florence O'Driscoll, I am prepared gladly to believe that this is true. But I was once persuaded to taste a South African Champagne and for sheer horror it would be hard to beat it. Take some pine sawdust, soak it in water in which you have previously soaked some grapes, add a little alcohol, and you will have a liquid fit to compare to the stuff of which I speak. A South African Sherry seemed to me little worthier of attention. I do not know if Constantia is now produced; but I fear that I shall be cautious about it. South Africa should content herself with the admirable eating grapes and other fruits she sends us, without seeking to compete in a market for which she would appear to be ill-equipped. [CAMPBELL: *Maurice was too prejudiced against Empire or Dominion wines to display any interest in them and his anathemas must not be taken too seriously. His legal and dialectical mind resented the nomenclature adopted by both Australia and South Africa and led him himself to adopt an attitude of 'can there any good thing come out of Nazareth?': a well-proven error.*]

Of Australian wines I can hardly bring myself to write with patience. On the Orient Line they used to stock a modest little Claret of Australian origin which was completely unpretentious, and quite equal to the duties of the *ordinaire* that it professed to be. And Tom Huxley once gave me a few bottles of so-called Australian Burgundy, which he told me was ten years old; and it had something of a Burgundy flavour about it, though not good Burgundy at that. With these two exceptions all Australian wine that has come my way has been detestable; and its inferior nature was not rendered less obvious by the way in which it borrowed the phrases in which George Meredith had praised genuine Burgundy to advertise itself. The Australian soil appears to be so rich that it cannot produce a wine. It is stiff with iron; and the so-called wine almost goes rusty in your mouth. The grape must be of a very coarse kind; if it is a French vine that has been carried out to Australia, it has degenerated in the new soil. Its produce is horrible; hot, coarse and flavourless, it puzzles me how people can continue to drink it. I suppose that for some time to come it will be the only beverage wine that we shall get; if so, advertisement should be unnecessary. But when once more French wines come to set a crown upon the table of the victor let Australia content herself with her many glories in respect of which she stands unchallenged, and forbear to steal the language applied to French wine falsely to insinuate that it was written about her own inferior products.

BRANDY

BRANDY is the soul of wine, and Whiskey is the soul of beer. The distiller takes the one substance or the other, and by vaporizing them and capturing the result he produces a new liquid, unrecognizable as akin to those from which it has been distilled. Brandy can be made from any wine; but although Cognac is Brandy, Brandy is not necessarily Cognac. Cognac is a strictly delimited part of the Charente in which are grown grapes so sour as to be uneatable, but capable of making a wine which is just as undrinkable. But then the distiller comes along, and from that despised wine he extracts a Brandy that is sometimes the greatest and most delectable of spirits, and is never insignificant. In Armagnac there is another district with the same chalky soil, where the same detestable grapes make the same miserable wine; and from this also is distilled a Brandy of great distinction, although not quite up to the standard of Cognac. There is no other district in the world that produces Brandy of quite this elegance; but the Spanish Brandy made of Xeres wine has great merit. [CAMPBELL: *An exaggeration, due, I fancy, to the spirit of friendship.*] In Burgundy and in Champagne they throw water on the murk or pressed grapes after the vintage, and distil from this a liquor called *marc*, beloved of all Frenchmen but not by me, even though I have occasionally had a good glass. The Germans make their own Brentwein, and they are welcome to it; it is coarse and harsh, and is well suited to their rulers if not to themselves. I have never tasted Portuguese Brandy, nor, indeed, that of any other country. Mine is a Gallic loyalty.

I once asked the late lamented Oddenino to advise me how to procure some good Brandy. He smiled. It is fair to say that the smile was a trifle sardonic. 'You want to buy some good Brandy?' he said. 'It is easy. First, you go out to Cognac in a good vintage year. Then you find a reliable *vigneron*. Then you buy your Brandy from him. Then you keep it in cask for twenty years. Then you bottle it. Oh, it is easy. Of course, you will be careful to choose a good vintage year; and it is most important that you should choose an honest *vigneron*. There will be others looking for him, too. Indeed, Mr. Healy, it might be better for you to go to a good wine-merchant.' He was a wicked old man, Oddenino; in the Pickwickian sense, of course. And both branches of his advice were excellent. How often do we hear a wonderful story of a friend who in the course of a visit to Cognac found a farmer with a little keg of Brandy that he wanted to sell, too small to be worthy of the attention of the great merchants? I suppose that this is by now an organized industry, like the sale of Genuine Antiques in the Shakespeare country. Beware of such bargains. There *are* such to be had; but there are some very acute noses constantly on the trail, hallooed on from the bottom of Saint James's Street and other places; and it would need an uncanny astuteness on the part of an amateur to succeed in gleaning from a field reaped by the big houses and raked by the scouts of the leading wine-merchants an overlooked little cask of good Brandy.

That such things do happen, let this narration bear witness. In 1925 I happened to be dining one night at the Isola Bella Restaurant when the second Signor Micotti, since left, asked me would I take a Brandy with my coffee. Now, Soho is not the likeliest place to find a good Brandy, so I laughingly said that I thought I would prefer a Strega. But Mr. Micotti was persistent, and my refusals could not be carried further without rudeness, which I should never have offered to so good a friend. So, resignedly, I asked what this wonderful Brandy was. He told me that it was a little parcel he had recently acquired that was supposed to be very old and fine. 'Like every Brandy offered in a restaurant,' I said to myself, and watched the bottle being opened. It was a bottle of Burgundy shape, and its newness shone from its surface; it was deeply sealed with dark green sealing-wax, and under the wax were crossed ribbons over the cork, which was almost white, and could hardly have been driven home more than a month before. The label looked equally new; it was printed in the style affected by French printers of the period of the Directory. But with brazen assurance it proclaimed the contents of the bottle to be 'Pure Cognac: 1805.' That was the year of Austerlitz; and the shoulder of the bottle had a medallion blown on it, containing an embossed head and shoulders of the Emperor, which was repeated on the label! A more audacious joke I had never come across, and I smiled indulgently as a little of the contents were poured into my glass. I lifted it to my nose, then I suddenly set it down. 'Mr. Micotti,' I said, 'can you get me any of this?' He shrugged his shoulders and said: 'Yes, Mr. Healy, at a price.' 'Well,' I said, 'I do not know what I am letting myself in for, but could I have three bottles?' He bowed. 'I think it will cost you two pounds a bottle,' he said. I expected him to say five, and I gladly clinched the bargain, without too closely scanning the provisions of the licensing laws. And I resumed the consideration of my glass.

Of course, assuming that a really old Brandy were found, either in wood or even in undistinguished-looking bottles, there would be no crime or fault or harm in providing it with a new residence more in consonance with its merit. The very newness of the cork and glass were calculated to catch the eye of the purchaser of this bottle. I know a Brandy, not entirely devoid of merit, which insists on arriving in a bottle which the vendor had at some time thinly coated with gum and then put away in the dustiest place he could think of, to acquire the appearance of age. That is a form of treatment far better calculated to deceive the ignorant; and I am sure that the average purchaser of such a bottle believes that the contents have been maturing for years inside, unconscious of the fact that spirits do not alter their character in glass. So, obviously, this Brandy was not to be condemned as a fraud until it had been tasted.

At that time Messieurs Christopher were supplying me with a Brandy that came from the vineyard of Monsieur Gabloteau and had been distilled in the year 1893: a good Cognac year. A few evenings later it was possible to compare this with the Austerlitz Brandy; and I can best convey the result by saying that the younger Brandy tasted like crisp, honest spring water, but the veteran tasted like cream. Those who knew most about the subject almost beat me for doubting its authenticity. It had mellowed, in their opinion, for a great deal more than twenty years in cask; it had shed all the hoity-toityness of youth, but had preserved its vigour. Its flavour was a softened cleanness, what is sometimes unsatisfactorily called nutty. It was a glorious thing. When I went on my holiday a few days later I left two and three-quarter bottles of it in the cellarette of the sideboard. I left by a night train; my mother returned home by the morning mail. She was met on the doorstep by her cook, who desired to give notice. Pressed for a reason, she said that she had drunk all that was in the cellarette and did not feel very well. As she showed no signs of alcoholism, and as there were a couple of bottles of Whiskey and some Vermouth in the cellarette as well, I incline to the view that she had sold them. I hope so; they might have reached a more worthy throat. *Sic transit gloria mundi.*

A few months later Mr. Micotti told me that he could get me an even better Brandy from the same source. But this time the gods were inclined to the view that I had had my chance. The resulting 'bargain' owed more to the plum tree than to the vine, and I will not swear that almonds had not an influence as well. I should have known better.

Let us begin our consideration of Brandy by taking the highest as our exemplar; if we study Cognac the rest will be added to us. Cognac, then, comes from a district of the Charente in the vicinity of the town of Cognac, where the soil is tremendously chalky. This area is classified under five headings. First, there is the best part of all, called Grande Champagne; the next best is called Petite Champagne; then comes the Borderies; and finally the Fins Bois and the Bons Bois. It is a pity that the fourth division should be called Fins Bois, because that gives the word 'fin' an official status in the nomenclature of Brandy, and is certain to be confused by the ignorant with the word 'fine,' which has no official meaning. 'Grande Fine Champagne' is a phrase without any precision, and I suspect that almost anything could safely be sold under such a label, because the Grande does not directly qualify the 'Champagne' (and 'Grande Champagne' is a legal indication of origin); it qualifies 'Fine Champagne,' and there is no such area in Cognac. But the sale of Brandy by the tot invariably involves the description 'fine'; a 'fine maison, sans blague' used to be the method by which a certain connoisseur used always to put a waiter on his honour. Now, a 'fine' means everything, or nothing, according to the individual honesty of the vendor. It ought to mean that the Brandy is pure and good; it often means that the proprietor of the particular restaurant has bought something without pedigree, which he cannot label with precision.

So much for this digression; let us get back to Cognac. In the chalky soil that I have described there are planted vines of the species called the Folle Blanche, and they, like the cider-apple tree that grows an uneatable apple from which good drink is made, produce the grapes that make the undrinkable wine from which Cognac is distilled. There is nothing about the vintage to call for comment, and the very careless guide might sum up the distillation by saying that the wine was twice 'boiled,' the first process extracting a product divisible into head, tail and heart, and the second getting rid of the head and tail. But like most modern condensations, this would give a very imperfect appreciation of what takes place. In the first place, let it be clear that in Cognac the patent still, capable of being heated to the most intense degree, and thus capable of extracting alcohol from almost anything, is not known. Only the good old-fashioned pot-still is used; and although in some of the great establishments there may be, for all I know, stills of vast proportions, one of a capacity of about 2,000 litres is the most usual. Into this the wine is put; it is carefully heated for about a working day, at the end of which period about a third of its contents have vaporized and passed along the worm into another receptacle. This distilled liquor is far from pure; a second, rather longer, and more careful distillation now takes place; there comes forth first a very alcoholic portion, called the 'tête,' which is still unfit for consumption; it is followed by the 'cœur,' which is real Brandy, and forms the bulk of the contents of the still; it continues to flow until the alcoholic strength has dropped below a certain point; then the rest is gathered elsewhere as the 'queue.' The 'tête' and the 'queue' go back into the still with the next lot of wine.

Those interested in following up this process in greater and more scientific detail will find a full analysis of the making of Brandy in Mr. H. Warner Allen's book, *The Romance of Wine*, from which I have learned much of what I know. But it should be mentioned that the foregoing description only gives the essence of the process, which varies in detail, and in the degree of care and attention, in different localities. Mr. Warner Allen hints at a danger awaiting the amateur Brandy-seeker; the small proprietors frequently make use of the services of a travelling still, which peregrinates from farm to farm and earns its living by hiring itself out to the owners. Naturally, the more rapidly the wine can be distilled the sooner will the still be able to move on to another job; and accordingly Brandy so distilled is rarely worth the attention of a connoisseur.

Once the spirit has been distilled nothing remains in the case of a 'straight' Brandy except maturation in cask. The casking of Brandy is a matter of great importance; oak is your only wood for maturing either wine or spirit, and the Cognac casks are made of Limousin oak. The men on whose forehead God set the mark that caused them to be invited to taste Walter Berry's Gabloteau 1906 will remember its almost unearthly pallor and its correspondingly ethereal bouquet and flavour. Monsieur Gabloteau chose every stave of every cask into which that Brandy was poured; and nothing less than twenty years old was permitted to be used in their manufacture. At that age, with proper seasoning (which does not imply the addition of pepper or salt or anything else) the timber has ceased to give out anything in the way of colour or flavour; the obscuration of that Brandy was .00-something. If Brandy could be matured in a substance like glass it would reach the consumer as colourless as water. But Brandy cannot breathe through glass and must therefore take up its dwelling in wood; and from even the most carefully matured wood a little colour will come. It ought to be the only source of a Brandy's colour; but I am sorry to say that various mixtures are used to give a deep brown to the liquor. 'Caramel' is the commonest; it consists of brown sugar, sometimes flavoured with prune-juice or other substances. Take a little Brandy (it is enough to dip your fingers in the glass) and place it on the palms of your hands; then rub them together until the liquid has evaporated, and smell your hands. If there has been an adventitious aid to the flavour you will detect the odour at once. But in fairness to the Brandy, try the experiment a second time; on the first occasion the soap with which you washed your hands may be asserting itself.

Now Brandy is doctored in this way because the Brandy-drinking public look for strength of flavour rather than delicacy; and little by little there has grown up a belief that Brandy browns with age. I once met a most eminent expert in Brandies, whose services had been borrowed by the French Government in days when it took more interest in Cognac than in Vichy to advise them about the industry. And I could not shake him in his view that Brandy browned with age. It is admitted that white wines do; but I think most other things lose colour in ageing. And if the Pope of Rome (who, be it remembered, claims no infallibility in his opinion about Brandy; indeed, I doubt that the poor man permits himself many opportunities of tasting it), if the Pope of Rome, I say, were to assert that Brandy grew darker with age I would be as obstinate as Martin Luther in contradicting him.

But this heresy has had its effect in Cognac, where, after all, the good spirit is a means of livelihood before it is a work of art; and the great shippers have in many cases decided to pander to the taste so expressed, and, by using their skill as blenders, give the public Brandies which will always be of the same character and quality. 'If this be a virtue in Xeres,' they say, 'why should it be a vice in Cognac?' And so they have prepared in their establishments great reserves of Brandies of various vintages, by the judicious mixture of which they claim to be able to serve up to the public a constant article of excellence and worth. Furthermore, there is another legend, with a substratum of truth in it, that I think must be carefully fostered by certain sections of the trade. The form in which I first heard it was, that if a cask of 1811 Brandy had got down to its last tenth, and was then refilled with a Brandy of a younger but sympathetic vintage, say 1834, the cask became within a few days a full cask of 1811, the older spirit having endowed the younger with its quality, while receiving the strength and virility of the other. The practice of refreshing a cask is old and recognized; but the resuscitation is not as accurate a re-creation as this story would suggest. But one house at least has adopted it as a commercial principle. Messieurs Bisquit, Dubouché until recently were wont, and for all I know continue, to advertise their Brandies, offering to supply apparently unlimited quantities of their 1865, their 1834 and even their 1811. After 129 years there could hardly be more than a tiny supply of an unrefreshed 1811; and I am sure that they do not wish their advertisement to be read as anything but an offer to supply a Brandy blended to match the original 1811. In these days of spiritualistic credulity what a sensation C. B. Cochran would make if he were to advertise a revival of *The Bells* with Henry Irving in his original part!

But this practice of some of the houses in Cognac has caused a general want of faith in the authenticity of a date-label on a bottle of Brandy. Take 1865. I am told that 1865 was not a vintage year in Cognac at all; 1864 was the great year, as in Bordeaux. But in 1865 one of the most enterprising of the modern houses began business; thirty years later it made such a *réclame* for its 1865 (which was in fact very good) that it has gone on selling it ever since, and has compelled a lot of rivals to sell an 1865 also. Napoleon III brought a great solace to the consciences of many shippers, who were able to speak of Napoleon Brandies with less mendacity than would have been possible a year or two before his accession to the imperial throne.

However, all this blending and boosting has resulted in a very serious step being taken by some of the biggest and best houses in Cognac. They have decided to focus their attention upon certain marks and to blend these marks to a constant standard. If ever you are in a place where you do not know the wine-list, you can order with confidence your Hennessy XO, or your Martell Cordon Bleu, or your Exshaw 40 Years Old; you will always get the same Brandy, and it will be a completely honest, albeit a blended, Brandy. But if you can get a bottle of Martell 1904 (of a reliable bottling) and compare it with the Cordon Bleu (which I think is now about the same price) you will see the whole matter in two tastes. It may well be that you will prefer the Cordon Bleu (into the making of which I believe practically no Brandy younger than forty years old has gone); but I shall continue to prefer the delicacy of the 'straight' 1904. Now, I have it from the horse's mouth (if this be not too rude a way to speak of a member of that distinguished family) that the Hennessys themselves always drink their XO as the family liqueur; and I myself always enjoy an XO or a Cordon Bleu when I taste it. But, just as the best of ready-made suits will never fit quite as satisfactorily as one made to measure, so do I find that a standard blend lacks the finesse and the individual charm of a genuine vintage Brandy.

Are the 'standard' Brandies pure? Yes, subject to this. Some palates cannot agree to the dryness of an absolutely pure Brandy; and in Cognac (and no doubt elsewhere) there is a custom of the trade to add the least taste of a syrup made of cane sugar dissolved in Brandy. In a 'faked' Brandy this dose is often increased; but in the standard 'marks' it is almost imperceptible, and if a fault at all, is venial. Even some of the 'straight' vintage Brandies receive a tiny dosage at the time of shipping; it may be that some of them get a further drop when being bottled, which shows the need of care in discriminating between different bottlers of the same vintage. I have known the Martell 1904 as bottled by J. H. & J. Brooke, Avery & Son, the Army & Navy Stores, John Gardiner & Sons, Henekeys, the Newcastle Breweries and one or two others. I shall not express my opinion as to the best and worst, but there was an enormous range of difference, which was reflected in the price as well as in the flavour.

Brandy is usually landed in cask in this country a year after the vintage; it matures in wood in bonded cellars, losing a considerable amount in bulk when doing so. The London Dock cellars are supposed to be very favourable to the maturation of Brandy, having an exact sympathy of moisture in the atmosphere, so that the loss of bulk is not as serious as elsewhere in dryer cellars. After about twenty years bottling takes place, after which the quality of the Brandy does not alter.

Although I have only mentioned four shippers so far, they do not exhaust my allegiance to Cognac. I have always found

Hine a name to bless; Delamain is very reliable, and there are many others to be esteemed. I express no opinion about the well-known brands of Otard and Courvoisier, because my meetings with them have been rare; and I have such a habit of getting names mixed that I hesitate to praise what I have met rarely lest I should unconsciously substitute another name unworthy of compliment. Besides, Brandy is an article in which you are inclined to stick to what you know. When I began, Christopher's 1893 (the Gabloteau) was still plentiful; and I stuck to it until Henry Brooke introduced me to the Martell 1904, which I subsequently got also from Ronald Avery. The Hine 1906, and the Martell and Hennessy of the same year have been favourites of mine; and of course that wonderful Gabloteau 1906 was a revelation. I met the Gabloteau of 1911 in Leicester at a modest price and found it excellent, and I have drunk one excellent 1913, supplied by my friend, Charles Roope. A 1922 from Chalié, Richards is the youngest vintage I have drunk; thanks to its comparative youth, it was modest in price, but it was quite mature enough to be enjoyable.

Three-Star Brandy is an excellent, straightforward product that goes well with soda-water as a long drink, or helps to make any of the Brandy cocktails; but it is an error of taste to serve it as a liqueur. There is an even cheaper variety of Brandy that is good enough for the kitchen. Most shippers have marks of this kind.

Armagnac is the younger sister of Cognac, and a good Armagnac is far better than an inferior Cognac. For us it suffers from the fact that it is not made near the sea or a river; so that transport to England is difficult, and as a result the trade was never a direct one, and we are not familiar with the names of the shippers. As a result, I should hesitate to affirm that I have ever tasted an Armagnac of which it could safely be said that it was a pure and genuine product. The French are incorrigible when it comes to doctoring their spirits; and as I do not think I have ever come across an Armagnac that was landed in England like a Cognac within a year of its vintage, I should be sorry to pledge my oath that it had never been 'improved' before crossing the Channel. The best samples I have met have been clean and pebbly; they have seemed to me 'thinner' (not in the sense of being weaker) than Cognac and rather dryer. But I am no expert, and I can give no advice of value.

Spanish Brandy I know chiefly through Messieurs Gonzales, Byass, who modestly underrate its merit. I do not pretend that it ought to be compared with Cognac, but it can round off a meal very agreeably, and it has all the appeal of a simple honesty. It will probably be cheaper than Cognac for some time to come, so it may perform a gracious duty as a deputy.

Although Brandy can only be made from wine, there is another spirit that deserves a mention here, which is distilled from Cider and is known in England as 'Applejack' and in France as 'Calvados.' I once enjoyed an English Applejack that was eighty years old; a week later in Paris I was given a Calvados of comparable age. In neither case did the spirit agree with my digestion, so that I will leave it to those who like it to appraise its merits.

To go back to Brandy, I marvel at the badness of the German product. I suppose that the wines are too fruity; for even the red wines of the Rhine have a volume of fruity flavour that is unmatched elsewhere. And all the good Brandies appear to be made from undrinkable wines. As a result, any German Brandy that I ever tasted had a flavour that carried a suggestion of a mixture of cigar-smoke and burnt sugar. Awful stuff! For Brandy ought in my opinion never to be a heavy liquor; and Germans like most things to be heavy, although their fingers and their beer are exceptions.

WHISKEY AND OTHER SPIRITS

TO follow a word on Brandy with one on Whiskey is not so irrelevant as at first it may seem. More than thirty years ago when I was supposed to be a student I enjoyed the hospitality of my uncle's house at Chapelizod, whence Iseult cast her spell over Tristan. The River Liffey ran over a weir at the foot of the garden, and flowed on to drive the incidental machinery of a distillery a half-mile lower down. Day and night the machinery clanked and groaned; what it was doing I do not know, for I should have thought that distillation was not a mechanical process. But to this day I do not know the name of the distillers; and, although their transport reduced the road between Chapelizod and Islandbridge to a repository of dust, I have my uncle's authority for the statement that not a drop of Whiskey was ever sold in Ireland. It was all shipped over to Cognac. [CAMPBELL: *A brave statement! did that account for the Irish clanking and groaning?*]

Now do not smile at this story until you have heard another one, which I told to the victim himself, who only laughed and did not contradict it. Many years ago Monsieur Hennessy was visiting Cork, and was entertained by the hospitable Dominic Daly, one of our leading citizens, and one also engaged in the wine and spirit trade. After lunch the host said: 'You would like a liqueur with your coffee?' Monsieur Hennessy concealed a grimace, and accepted. 'Bring that peculiar decanter,' said Dommy Daly to his butler. A few minutes later he turned to his guest and said: 'Now, tell me what that is.' 'Oh!' said Monsieur Hennessy, 'it is very difficult to be precise in these matters. I should call it a Grande Champagne: probably an 1893.' 'It's John

Jameson's Ten Year Old Whiskey,' replied Dommy. [CAMPBELL: *See page 36 and my note thereon.*]

I have myself repeated this trick on many occasions, and the only guest who did not guess a Brandy suggested a Vieux Marc—that was a Seven Year Old JJ. Oddly enough, the only time that the trick was attempted on me the tempter was Ronald Avery, and he was dishonest enough to tell me that the potion was Martell 1884. I hesitated. I said: 'If you had told me that this was a Vieux Marc of 1884—it's an old Irish Whiskey! an old Irish Whiskey!' And it was!

Now, I think that there is something about Whiskey which is completely 'other' to the taste of Brandy. I am sorry to call it a sickliness of flavour; but I do find Whiskey mawkish to the taste when compared with Brandy. And I do not think that I should ever mistake a Whiskey for a Brandy, no matter how good the Whiskey was. I can remember a Scotch Whiskey, bottled in magnums by Christopher's, which used to make a good Christmas present; I used it to make my peace with Bill Bannerman, may God be good to him, on more than one occasion when I had driven him to desperation by my vagaries on the Wine Committee of the Windham Club. [CAMPBELL: *Yes, I am afraid 'vagaries' were to be expected on the wine committee, but his invocation to Heaven on behalf of his colleague would compensate for them all.*] And, by the way, poor Bill was responsible for one of the best Brandy ramps of which I have ever heard. Hew Brown in those days *was* Tanqueray's; he owned the business. And he told me that Bill had offered the Club a magnificent old Brandy, of undoubted purity and age, which was rejected by the Committee because it was not full enough or dark enough. So Bill, who had rashly committed himself about the purchase, asked Hew what he had better do. And Hew, still having the spirit of a Harrow boy, said: 'That's all right, let me have it.' And he turned it all out into a mixing vat, to which he added this and that; and he rebottled it and sent it back to Bill. At the next wine-committee meeting Bill said: 'I have a remarkable sample of Brandy here'; and to cut a long story short, the wine list of the Club showed a Brandy which it called 'Tanqueray 1865' for many years afterwards, until I insisted upon a correction. To do Hew justice, he had tried to get it corrected also, but in vain. The implacable Bill insisted: 'They would not take the genuine article I offered them; they took your beastly fake; let them stew in it!' I am bound to say that the fake was by no means a beastly thing; but as it cost 2s. 6d. a glass as against 1s. 6d. for the Martell 1904 it was not ideal value. But God rest the soul of old Bill Bannerman; he ceased to be Secretary because of his uncurable passion for vingt-et-un, which he insisted on playing after a special warning from the Chairman; and he died away from us all, in Switzerland, and his daughter sent his ashes home, so that we never had the opportunity that a funeral affords to pay tribute to the memory of a friend. Dear Bill! It was from him that I learned the precious truth that there is nothing like fish to give a smack to beer. I learned other things also, but I am not going to write about those.

What on earth were we talking about? Oh, yes; the flavour of Whiskey compared with that of Brandy. Well, Bill, peace to his ashes, was a good judge of both, yet I, writing with this assurance, have to confess that I cannot abide Whiskey, although I live in hopes; I have on one or two occasions tried it as a liqueur; but the Whiskey-and-Soda has never appealed to me, and I believe that I have not drunk half a dozen in my life. Whiskey-and-Soda at once suggests Scotch Whiskey; he would be a fool who would attempt to taste Irish Whiskey with anything but soft water, from the neighbourhood of a bog, if possible. I have tried to procure what a Scotsman would recognize as the best of Scotch Whiskey. I asked my dear friend, Peter Byrne, to get his merchant, or should I say in Scotland his factor, to send me a sample of the best of Scotland's produce. He riposted with a bottle of 'Talisker'; and I gave it the trial that I think any great produce should enjoy: I drank a little of it as liqueur. And I freely admit that it is a great Whiskey; and I should be sorry to insult it with Soda-water, or even soft water, for at a sufficient age I think it would make a very good liqueur indeed. But your well-known Scotch Whiskeys are not to be treated thus, nor are they meant to be. They are meant for the long drink, and the man who partakes of such must get a thrill from the knowledge that no less than forty or more distilleries have contributed to his pleasure. For your well-known Scotch Whiskey lays no fewer distilleries under contribution; these are blended into three or four main blends in Scotland and are then sent south in that form, to be combined into the final shape in London or Birmingham. They are admirable spirits for their function, which is that of dealing with a thirst calling for a long drink. I have never ascertained what proportion of the Scottish distilleries are pot-still, and what proportion are patent-still. For the chemist has been abroad in the Scotch Whiskey trade. The pot still is incapable of working above a certain temperature; the patent still can stand a much higher temperature. In its maw not merely the potato, but even timber, can be persuaded to give up its alcohol. When the wisdom of Parliament first addressed itself to the great question: 'What is Whiskey?' it appointed a committee presided over by Doctor Playfair, before whom a very knowledgeable scientist gave evidence. To him Doctor Playfair addressed the question, 'Has not science established that it is impossible by any chemical test to draw a distinction between Whiskey made naturally (which at the time of the inquiry meant by a pot still) and that made artificially?' The witness agreed. 'Then how do you dare to suggest that there is a difference between the two products?' 'I have an infallible test of my own,' replied the witness. 'And what might that be?' asked the scornful Doctor Playfair. 'The state of my head the next morning,' replied the witness.

Now, I do not wish to suggest that in any of the well-known proprietary brands of Scotch Whiskey there is present any formidable quantity of silent spirit; I only mean to say that in Ireland such a thing is practically impossible, for all Irish Whiskey with the exception of Dunville's is made in a pot still. You can drink your 'Paddy Flaherty' (The Cork Distilleries Company), your John Jameson, your Power, your Persse, your Bushmills, or any other Irish Whiskey, with the knowledge that nothing but malted barley has gone to make the mash from which it was distilled. And the motto that John Jameson has made familiar to all travellers, 'Not a drop is sold Till it's seven years old,' is equally true of the others. Not everybody likes the flavour of Irish Whiskey when it is first met; but it is an acquired taste, and I rarely find that any of my guests who have given a trial to my Irish Whiskey want to revert to Scotch. Of course I cannot always offer them the alternative of Talisker! And Professor Saintsbury spoke of Smith's Glenlivet as I have been speaking of Talisker. I never came across it.

Gin is a spirit that has been very slow in its advance in my favour. I remember an evening in the autumn of 1916 when in the Officers' Club at Poperinghe, I, determined to be a dog, called for a Gin-and-Bitters before dinner; and I could not get it down. I continued to find the flavour of Gin unattractive until I first married it to its preordained spouse, the juice of the lime; and I confess that a 'Gimlet' is one of the most attractive drinks, between long and short, that I know. (You mix one-third Gin, one-third Lime juice, and one-third Soda or plain water.) A John Collins is also an agreeable drink; it is just a lemon-squash with a dollop of Gin. There are agreeable Gin cocktails; but even my warm affection and admiration for the Navy has not totally converted me to Gin. It has the merit of being a strong diuretic; and in conjunction with sloes it can make a most agreeable liqueur. (Each sloe must be pricked, and the soaking must go on for months.) Both Charles Walter Berry and Barry Neame give their blessing to Sloe Gin, the latter in the practical form of making it; and I cannot persuade him to give me a bottle, he who is so profuse with his other gifts!

Gin is to be found in three varieties: Hollands, London and Plymouth. It owes its name to the fact that originally the Hollands variety was freely flavoured with the juice of the juniper-berry; Genièvre quickly got corrupted and shortened to Gin. Primarily distilled from malted grain, it was for long the subject of much corruption; turpentine is supposed to play a part in its concoction. But since the manufacture has been brought into a few and reputable hands this adulteration has been largely banished; certainly one may now rely on any of the well-known brands as being safe from deleterious matter. It differs from other spirits from being all the better for being drunk young. Hollands I have only once tasted; I thought it mawkish and horrible. London, exemplified by Gordon's, Booth's and Seager's, is straightforward in flavour; Coates' Plymouth Gin, now the only brand of this variety, is a Naval favourite. Booth's mature their Gin in casks that give it a straw colour; others are white.

Rum is a noble spirit; and if it were only for the comfort that it brought to the troops in the last war (in spite of the wretched fanatics who sought to deprive them of it) it would always have an honoured place in my affections. For a long time I disliked the flavour of Rum; then grew to like it immensely. It is made from material obtained in the manufacture of cane sugar; the best is produced by the skimmings of the pan, and the worst from molasses. The sugar-product is distilled; and sometimes pineapples or guavas are thrown into the distilling mixture. There is a difference between the methods of distilling Jamaica Rum and Demerara Rum; but I am at a loss to explain it. 'Bacardi,' a name familiar to every cocktail shaker, is not to be found in the Gazetteer; and I cannot aver that it is the name of a man, or the name of a place. All I know is that Bacardi Rum is as different from Jamaica Rum as is chalk from cheese. It is more delicate and refined; Jamaica Rum is essentially Rum-Bumptious.

As a tot, Rum is terrific. I hope there are many people who remember a scene between Leslie Henson and Sidney Howard in a musical farce called *Funny-Face* which had the misfortune to be put on in the autumn of 1928. First the gas blew up and isolated the theatre; then when it had found a new home the King fell dangerously ill, and damped the zeal of theatre-goers. But there was an ecstatic moment in the play when Sidney Howard, disguised as a butler, wheeled in a buffet, on which was a silver bowl of punch, and, by tapping it discreetly, attracted Leslie Henson's attention. Leslie Henson was persuaded (without difficulty) to partake of several glasses; and as each glass vanished down his throat his face recorded the effect of an earthquake followed by smiling sunshine. I think that the bowl probably was meant to contain neat Jamaica Rum. A tot goes down your throat like an incendiary bomb; there is a moment's confusion below, as the charge explodes; then there spreads through the system a strong desire to gather all your enemies and give them ten thousand pounds apiece. And next morning you wake up, still feeling a little bit like the mouse who had a drop and said 'Where's that damned cat?'

Rum also demands the hand of the lime-fruit in matrimony; obviously the one was made for the other. At the British Empire Exhibition of 1924 the Jamaica Pavilion offered the weary traveller a 'Planter's Cocktail,' which went to the recipe of 'One of sour, two of sweet; three of weak and four of strong,' in which the sweet was sugar, the sour was lime-juice, the weak was water and the strong was Jamaica Rum. It was an admirable restorative for a tired man; but I seem to remember that few of the refreshed pilgrims showed any anxiety to renew their perambulations.

I have never drunk Saki or any of the rice preparations; nor yet have I tried any of the spirits of Africa or Asia. Swedish Punch is a compound in which Rum plays its part; and Schnapps is only a form of Gin. Liqueurs are merely sugar syrup charged with alcohol and flavoured according to the fancy of the maker; any doctor can make his own, and some of the most famous can be imitated at home without loss of enjoyment. This therefore finishes the chronicle of all except two of the most popular and widespread drinks, to which we may give another chapter.

BEER AND CIDER

BEER is the most English drink of all. It takes many forms; in Germany they prefer it in the shape of *lager;* in England, some prefer it mild and some prefer it bitter; some prefer draught and some prefer bottled. All are good. I consider that my education was sadly neglected, in that I never tasted beer until the year 1927. I was ready for it; for I have always thought that the best Cheddar cheese is that which has a strong beery flavour. But whenever I was offered a glass of beer I sipped at it nervously, and was nearly sick. Then I found myself on one scorching summer's day walking down the main street of Cologne. All around me were shops and restaurants where smiling waiters were bringing to their customers tall beakers of ice-cold *lager*, with a frost forming on the outside of the glass. I could resist it no longer; I ordered one for myself; and when it arrived I *opened my throat and swilled it down*. And then I knew how good beer could be. Of course, the fault had been in myself; I had tried to drink beer the way one should drink wine: by sipping it. And there are few things in this life so revolting as sipped beer. But let it go down your throat 'as suds go down the sink,' and you will quickly realize that this is a true friend, to be admitted to your most secret counsels. Long draughts with an open throat are the secret.

Lager seems to owe its otherness to the method of its fermentation. When after the last war I was trying to get beer for the troops in Germany I got one letter which spoke of 'our by-a-special-process-top-and-bottom-fermented-beer.' I felt it must be good with a description like that; and I ordered a consignment. It *was* good. English beer, on the other hand, is apparently only fermented from below. Also, hops are supposed to play a large part in English beer; I do not think that *lager* contains hops, but it usually betrays the presence of more than a touch of garlic. I know of no scientific reason why *lager* should demand icing to be served in perfection, while English draught beer is undoubtedly harmed if its temperature is brought down to anything below what would be described as cool. Our beers are more potent, also; 'One over the eight' is a phrase to indicate drunkenness, whereas I have myself put away 31 litres of German *lager* in one twenty-four hours, without being conscious of any evil effect. I may add that this statement so shocked the editor of *The Listener* that he twice cut it out of my contributions to a teetotal controversy in the columns of his paper. But there is really nothing in it. The late Father Tom Finlay, one of the wisest of men, once told me that when he as a young Jesuit went to Munich to pursue his studies, the Master of Novices addressed the young arrivals in friendly warning. 'You will like our beer,' he said; 'and you will perhaps be tempted to drink more of it than is good for you, not knowing its powers. Well, I would counsel you to set yourself a limit, and not to exceed that. You may not know what limit to set yourselves; my own limit is 17 litres a day, if that will serve you as a guide.' So I think that a tourist on holiday at a more mature age need not be ashamed of having merely doubled this minimum.

Although I regularly drink *lager* in Germany (or used, when that country was fit to visit), I rarely drink English draught ale. This does not come from pride or want of appreciation of its excellence; there is a slightly soapy flavour that I meet too often in it. There are few of the London ales that I care for: Watney's, perhaps, coming first, although there is a house of call near Belgrave Square kept by an old friend of mine where I drink Fuller's with great pleasure. But I think that for really good beer in England you must go north of the Trent. And I do not refer to Burton; I have drunk delicious brews of Bass and Worthington and Ind, Coope; but I have also drunk them when they were not so good. Offilers of Derby I have always found agreeable and sometimes worthy of still higher praise; Stone's of Sheffield brew a steady, good ale; the John Smith (Tadcaster) Brewery I have also found very reliable; and it is a matter of regret to me that I have never had an opportunity of trying Lancashire ales, which presumably have helped to make the Lancashire Fusiliers what they are; and what higher praise could be given? Younger's is the only beer from north of the Border that I know, and I grudge it no compliment. Of course, I am at the moment speaking of draught and bottled ale; for Gordon's *lager* is in my opinion as good as any German competitor, if not better. The Wrexham *lager* is good also; but I confess that I do not like the taste of Barclay's. This need not worry the brewers, for I do not like the flavour of Guinness either; but I cannot believe that there could be a consensus of good opinion about an inferior product as there is about Guinness.

Beer ought to be given greater honour in England; it is an insufferable thing that certain restaurants and hotels should deny it a place at honourable tables. Take that most hospitable and elegant of countries, Sweden. Every dinner of ceremony begins with beer and goes on to wine. [CAMPBELL: *The Medical Profession in this country is almost unanimously opposed to a large quantity of liquid being taken at the beginning of a meal, and you cannot drink beer out of a wine-glass. Personally, I deprecate the consumption of any* apéritif *except wine before meals at which wine is going to be served, both on hygienic grounds and also to ensure proper appreciation of the wines to follow. Maurice would appear early in his next chapter to support my view.*] A tankard of ale ought to be a recognized decoration of the proudest banquet; and, by the way, ale ought to be drunk out of a tankard: glasses seem an effeminate dress for it. Perhaps we may take comfort from the fact that many of those who banished honest English beer from their tables are at the moment banished from those tables themselves; and in their absence a better tradition may grow up.

Cider is an honest and a pleasant drink. Its sweetness and fruitiness commends it to children; it was at the age of ten that I first got loose on the Cider barrel in Mr. Aherne's shop in Lismore, and 'liked it better than lemonade.' It must have been a sweet Blackwater Cider which I should probably find too sweet to-day. But I confess that Cider is one of the few products in which I have maintained a sweet tooth; I find the elegant Champagne-Ciders of to-day too dry and austere for me. And Cider is at its best in the form of Cider Cup.

Here is a recipe that has proved popular: Take, for six pints of cup, two oranges and a lemon, two largish apples, two bananas, about an inch of a good-sized cucumber, some grapes and some of any berries that may be in season. With a potato-peeler take off the thin rind of the oranges and lemon; put the rind at the bottom of the jug, and squeeze the juice and add it; slice the other fruit and set it with the rest in the jug to steep in the liquor to be added. Of course, there ought to be a sprig of borage; every cup needs that. Pour on the fruit two wineglassfuls of Brandy, a wineglassful of Sherry, a wineglassful of Cointreau, a wineglassful of Lime-juice and one of Kiaora Lemon-squash. Stir this mixture well with a wooden spoon, bruising the fruit; then let it stand in the refrigerator until the time when it is to be drunk. By its side set two quart flagons of Whiteway's Medium Sweet Cider. Add these just before serving, and stir again. This will bring comfort to the most heated head.

FLOTSAM AND JETSAM

DRUNKENNESS is an abominable thing, and those who regard it as a pleasure should be ranked with those fanatics who gash their limbs and sit on spikes. It is rarely intentional, however, and any advice that can help to avoid it will never be out of place. So the first hint I would give to the wine-bibber is, Never start drinking on an empty stomach. Almost the only criticism I have to make of George Saintsbury is that he contradicted the general opinion on this point, and held that drink had no more effect on the system when the stomach was empty than when it was full. He was in manifest error. What causes drunkenness? The effect of the alcohol and other components of the drink upon the nerves. If the nerves of the stomach are subjected to too violent an attack they will 'gang all agley,' and will communicate their infirmity to those that terminate near them in the brain. Sometimes it is the other way about; the nerves of the semi-circular canals get irritated by agitation of one sort or another (as, for example, a channel crossing), and their unsettlement upsets the stomach nerves, with the result that the patient becomes violently sick. In the reverse direction an unsettlement of the stomach nerves will cause the nerves of the semi-circular canals to lose control; and at once the patient begins to reel and stagger. Other nerves may be affected also; the co-ordination of hand and eye may be lost, or the co-ordination of the vision of one eye with that of the other, causing the patient to see double. The presence of pink rats and purple snakes is explained in a somewhat more recondite manner, but the principle is the same.

Now, if alcohol is allowed to wage a *blitzkrieg* on the stomach the effect will be the same as the recent attack on Belgium and Holland. The enemy will come up against a position to all intents and purposes undefended. If the stomach is empty, the patient is hungry, and the nerves are Fifth-Columnists, waiting to welcome any invader. But if the first onrush can be delayed a little, then Nature will build up defences against the attack; she will strengthen the power of resistance in the nerves; so that when at last they receive the full brunt of the onslaught they have been tuned up to meet it, and are in a position to stand up to it. In Roman days the gluttons used to precede their potations with a couple of spoonfuls of olive oil, just as exposed steel is sometimes covered with vaseline. A thin coating of oily substance in either case delays the attack; and when alcohol reaches an oily stomach it cannot immediately get at the nerves. I have known medical students and others to prepare for what the late Lord Curzon once called a Bee-ahno by adopting similar precautions. But you and I have done the same thing often without realizing it. What about the oil with which most *hors d'œuvre* are dressed? Or the soups, not so oily or greasy, but with a protective quality all the same? And the hot soup will bring blood to the walls of the stomach, which of itself will strengthen the nerves and improve their powers of resistance.

Now, if you are a wise man you will not have taken anything except a glass of Sherry or Chambéry before the soup (unless Madeira had been offered as an alternative to the Sherry). Also, you had observed that very sound rule of abstaining from all alcoholic drink until six o'clock had struck. Your inside had been resting since tea-time, and a little fillip, such as a glass of Sherry will give, would be just the thing to tone it up; to sound the Assembly and parade the troops ready for any emergency. If you had started drinking cocktails, in the first place for some unknown reason you would have been expected to drink several, whereas the conventions are satisfied with a single glass of Sherry; and several cocktails would in all probability have blunted the acuteness of the taste-nerves if the consequences had been no worse, and would thus have interfered with the enjoyment and appreciation of your dinner wines.

So we go in to dinner. Now, in this country it is unusual to see beer served at a dinner of ceremony; but the Swedes, to whose standard of elegance I have already paid tribute, have no scruple about beginning the meal with beer, and proceeding later to wine. But those mighty drinkers can also face a preface to the dinner proper, at which all sorts of delicious gobbets are washed down with *aquavit*, a powerful spirit of the Gin variety; and the ways of their northern climate may not be altogether suitable to our more tranquil airs. So let us assume that the usual rule is followed, and that nothing but wine is to be served.

At dinner, cut out the *hors d'œuvres*. Begin with soup. Both would be an *embarras de richesses;* and, as dinner is intended to be a substantial meal, you will want all your appetite later on. Let the soup be very good; and let it not be Mulligatawny, or any other over-spiced or flavoured product. For a wine dinner, a good *consommé* cannot be improved upon; for those who like a thick soup, a Crême Forestière will serve admirably, the flavour of the mushrooms making a good background for the wine that is to follow. A giblet or kidney soup can be very good, but it should have a glass of rather rich Sherry to qualify its richness. Vegetable soups are innocuous, and sometimes agreeable; pea soup is a favourite of mine, as is any soup made with potatoes. The Italian *minestrone* is excellent, but, like all these vegetable soups, it is very filling, and will not leave much appetite for the following courses. And at a wine dinner the function of the soup is not to satisfy; it is to warm and prepare. The richer the soup, the richer should be the Sherry served with it. Turtle soup, the aldermanic soup of the City of London, must never lack its glass of old Madeira. Fish soups seem to me out of place at a wine dinner; but a *bisque* of lobster can stand a fairly full Sherry, and nothing in the world can redeem any other fish soup. The serving of oyster soup is a felony, a sacrilege, a misprision of treason, a blow below the belt and, in short, table they are always served with the muscle unsevered and still attached to the shell. Personally, I think this a mistake; I can make but a poor hand of a job that her skilled operators can do much better). The only accompaniments should be a squeeze of lemon (some deny even this) and thin brown bread and butter. The tray of peppers and vinegars that usually comes round is a relic of barbarism, and has only an indirect purpose. To-day the admirable system of preparing the oyster for the London market renders him almost incapable of transmitting disease; but I suppose that a modicum of peril remains, and I always remind my friends that in the great oyster-poisoning case (was it from Weston-super-Mare?) only the teetotallers were poisoned. So comfort the generous creature in his last hours with a little drop of something, which will serve for a disinfectant also, for few if any bacteria can live in wine. Let it be either Champagne, Chablis, Meursault, Mosel or (if not at a wine dinner) Stout.

Baron Aäge Ancker once gave me a recipe for a prefatory *bonne-bouche* that I only once ventured to serve, and I do not want the Food Controller to hear about it. (It was many years ago!) You prepare a series of little beds of coarse brown bread and butter, about the size of half a crown; you scrape a powdering of red meat on these (personally I buttered the bread with butter into which a spot of Bovril had been worked); then on each you set a *couche* of caviar, and on the *couche* a small but fat oyster. A squeeze of lemon is again the only accompaniment indicated. The result is delicious, but it reminds me of Cleopatra's dissolved pearls, and I confess I regard it as something very near allied to gluttony!

Let us get on to the fish proper. At a wine dinner only the most discreet of fish must be served; it had better be a sole or a turbot, although John Dory can behave very modestly, and is delicious. For some reason the only place I know where he makes a regular appearance is A L'Ecu de France, in Jermyn Street, where with a sauce lightly caressed with taragon he is beyond the reach of mundane praise. Salmon may be eaten to the accompaniment of red wine; but the flavour of salmon is powerful, even when most delicious, and on the whole I think we will keep our salmon for another day; and the same applies even to trout. But the flat fish are innocuous to wine of any kind, and may be safely served with any accompaniment. If we are going on to drink seriously let us be content with a good white Burgundy, such as a Bâtard Montrachet of a good year; it will hold us, so that we shall look back to say farewell even after the Claret has arrived.

An entrée and a bird might be the next two courses in days without ration cards, or we might have a vegetable dish and a roast. I have a great weakness for such beans as *soissons* or *flageolets;* they always seem to me to make a good background for wine. The green vegetables do not seem to me to be so good, and personally I hate artichokes, which I am told are good for this purpose, although I cannot understand why, as they have a very pronounced and, to my mind, oily flavour.

Entrées will mean, nine times out of ten, baby lamb chops or something like that; if they are meant to precede a bird they can hardly be replaced by chicken; and sweetbreads are not a very good accompaniment to wine, nor are kidneys. With a dish of mutton or lamb or beef Claret may take its cue; a light Claret to begin with, a 1923 the ideal, or a 1933, if a younger wine be suggested. Such a wine might also go with a dish of white beans; but these, I think, can handle something bigger, and a good 1934 ought to suit them, or even a 1929.

If a roast is following it, stick to Claret and bring out your best, ignoring the rule propounded by that ignorant butler at Cana, who not only was not properly posted as to the state of his master's cellar, but also suggested that good wine should be drunk before the less good. (Probably a man from Blankley's, hired for the day.) Let the younger precede the older; and the junior growth the senior. If these rules clash, usually it is wiser to give precedence to the rule about the dates. A younger second growth before an older fifth growth, for example. This is not always so; to take an instance, I think a light growth like a 1923 should nearly always precede a big wine like a 1924 (if the 1924's are, as I think, only passing through a phase like that which seemed to knock out the 1920's a few years ago). But all these rules boil down to this: serve your least good wine first and keep the best until last, even though this should make pie of all I have been saying.

This is an appropriate moment to say a word about service. The good host will look after his wine himself, although the party may be too large for him to do the actual pouring. If he has to leave this duty to a Hebe or a Ganymede he should give careful instructions to them to watch all glasses carefully, not to allow them to become mixed, and not to fill them more than two-thirds full, refraining from refilling them until they are empty, or as near as makes no difference. Above all, do not let any of the series be removed from the table until the whole series has been dealt with; there is no objection to removing the Sherry glasses shortly after the soup has gone, or the white wine-glasses before the Claret is served; but if there are four Clarets, do not allow the empty glasses used for the first to be taken away until all four are ready for removal. Bouquet remains; sometimes a guest will desire to recall it when the wine itself has passed beyond recall. And let the servitors who are dealing with the solids be attentive and almost invisible—unheard, it goes without saying, nor should they give trouble to any other sense. It is of prime importance that a wine-taster should never be without bread [CAMPBELL: *This is very sound advice. Bread keeps the palate clean. Maurice has been wise also to mention only youngish wines with his dinner, many of which are still on restaurant lists or in wine merchants' bins*]; but it should reach his plate unobtrusively, without interrupting either his tasting or his talk. His plate should not be removed too hastily; such zeal may embarrass his neighbour, who has dental troubles that delay mastication. Nowadays, each guest nearly always has his own supply of salt, pepper and mustard, and in normal times butter; this saves a lot of trouble. The asparagus courses? They are never in the way, and may be useful. Ladies might use them to remove some of the superfluous make-up that, apparently, may be put on at the dinner-table in the course of the meal without any breach of the conventions of good manners.

When there are several wines of the same kind being shown, all should be shown at once; the first poured and tasted, then the second a minute or two later, and the others in succession at the same interval. This enables comparison to be made; in this case, comparisons *are* odorous. What has been written here about Claret applies without any alteration to Burgundy; but I think that Burgundy would probably be a better choice for a dish of game, which sets off its splendours with more sympathy than it would show to Bordeaux.

At a wine dinner there is rarely room for a sweet; but if there is a bottle of Yquem to show off, it needs a sweet accompaniment, of which the best is a nectarine. A plum is not quite fine enough; a good greengage is not bad, but the nectarine is ideal, and is better than any but the most perfect peach for this purpose. But if there be no Yquem, probably Port will be served; and cheese, or a cheese savoury, will prepare the palate for the change of wine. The cheese should be taken with the last of the Claret or Burgundy; Port does not care very much for cheese itself, and will be more appropriately flattered with a good Cox's Orange Pippin, or some other first-class apple. Once dear Edward Bunyard brought me a present of some Orleans Reinette apples that he had grown himself. They were the most delicious apples that I ever tasted; sweet, juicy, and crisp in the crunch. He told me that they are very hard to get; and although I begged shamelessly every year for a repetition of the favour, even his generosity was not able to grant my request. His trees more often than not failed him. Nuts, too, are supposed to high; but I am quite content with others that I have met. When I lived with my mother we had a little cook who used to buy a minute quantity of coffee-beans every day from Picard's, the forerunners of the modern popular Cona method of making coffee. These beans she used to roast and grind herself; and I have never tasted any coffee to compare with the result. But coffee, one would imagine, is one of those things which can always be made better on a small scale; and for this reason I was surprised on my first visit to the Hind's Head at Bray by the super-excellence of the coffee, which apparently was supplied in equal condition to every visitor. I like the coffee at the Isola Bella also; but most restaurant cups of coffee in London are things sent to try us. The whole secret seems to lie in the simple elements of a skilful choice of blend; a fresh roasting and grinding, and reasonable intelligence in the use of boiling water and powdered coffee. Yet the result frequently vies with the Frenchman's attempts at tea.

Now let me strike an original note. The repast is over; digestion has already commenced; our wines have done their duty, and our coffee has given a little fillip to a system perhaps tempted towards somnolence. Take at this stage a generous allowance of whatever form of mineral water best suits your system; and laugh at the doctors who threaten you with gout and indigestion and a host of other maladies. Some like Vichy-Celestins (not Vichy-Pétain); others prefer Contrexéville; others are content with Malvern Water which I am told is just honest H_2O. I used to rely on Périer at one time, and Apollinaris at another; I find a glass or two of Seltzer quite efficient nowadays; or even a good soda-water, like Schweppes, or Cantrell and Cochrane. This sluice to the system prevents such acids as may be in the wine from making your body their home; and it also removes thirst, which I find that wine does not do. So 'feast on wine and fast on water' and face the morrow with a confidence that you will not have a head or any other symptom of trouble. [CAMPBELL: *Every man must look after his own digestion, of course, but I cannot endorse Maurice's prescription. Brandy is possibly 'the best policeman,' if any policeman be necessary.*]

But you are not going to leave the table without a little

drop of Brandy, the best policeman to keep in order the cosmopolitan mixture of food and drink you have enjoyed. Now, for Heaven's sake, do not produce *ballon* glasses [CAMPBELL: *I agree about the* ballon *glass: it is an abomination and of no assistance to the brandy*]; they are mere ostentation, and are much more appropriate to the Soho Brandies that usually go into them than to any Brandy I would like you to stock. No doubt the bouquet of Brandy forms a large part of its charm; no doubt also it should have an adequate opportunity to develop and spread itself. But this is supplied by a Claret-glass, if it be of the appropriate shape; a slight pinching-in towards the lip, suggesting the tulip, is the ideal. And I think that even the Claret glass is too large, and I use my Sherry glasses for my Brandy. They are small enough to enable them to be held in the hollow of the palm, which will warm their contents; and there is quite enough space above the level of the liquor to enable the bouquet to develop.

Some people like liqueurs. I do not hold a strong view against them, but they call for comparison with Brandy, and the comparison must always be unfavourable to them. There was a time when Chartreuse took pride of place; but the unhappy adventures of the monks and the competition between the genuine liqueur made by the monks in exile and the spurious liqueur made in the original Chartreuse in the original vats by those who claimed to have successfully imitated the liqueur made by the monks caused a falling off in its popularity which may be restored now that the rivals have joined forces. Personally, I never greatly cared for the green variety, and I have hardly ever tasted the yellow. Benedictine would have been the principal competitor in the old days; I have always thought it sticky and exotic. Grand Marnier is more agreeable, and a great advance on the Curaçao on which it was based. I think Strega about the most pleasant of this group; but the modern Cointreau has outdistanced them all in my opinion. Kümmel, quite a different kind of liqueur, based upon the caraway seed, is agreeable and digestive. Crême de Menthe is also digestive, and very pretty to look at; but I find it a little insipid. Aurum is also pretty, and has an attractive flavour, besides enjoying a reputation that would link it with Tokay, and oysters and certain other articles of diet. The reputation is not necessarily well founded. Danziger Goldwasser is again pretty to look at, with its flecks of gold floating in the liquid; but it is a curio, not a comforter. Such things as Maraschino always seem to me to rank as flavouring essences rather than as liqueurs. Kirsch and Kirschwasser (which is made out of the fruit and which out of the stone I never can remember) never have appealed to me; and Slivovitz I have never drunk, I think. [CAMPBELL: *Maurice gives a wide choice, but the number of liqueurs is Legion, and even Eire is to-day trying to foist some of her own make on our markets. I rarely take a liqueur but, dining with M. de Sailly, the French Commercial Attaché, one night at L'Ecu de France, I was given a white liqueur, delicate, free of oiliness, with a fragrance and flavour of raspberries, called Framboise which I found fascinating, graceful and pleasant to sip. I never heard of it during Maurice's lifetime; I think he would have put it high on his list had he tasted it.*]

Glasses? Always should be fine and plain. Old Irish or Bristol glasses are a delight to the eye, but are not the best suited vehicles for wine. They are too heavy; the wine is a live thing, and longs for heat of your hand, which it can feel through a thin glass. And too much cutting distracts the eye from the natural beauty of the colour of the wine. And if this be true of all wine-glasses, why depart from the rule when taking liqueurs? Sometimes I am given a Brandy in an old cut-glass which is probably worth two or three sovereigns, but in which the unhappy spirit must lie uncherished and uncomforted by a sympathetic hand. The Brandy generally proves to be inferior. There used to be an exception made to this rule in the days when German wines were always served in green glasses; this was only to conceal the flecks of white which, apparently were in those days than in ours stem of which is hollow, resulting in a perpetual fountain of bubbles in the centre of the wine. In the days when I could still afford Champagne, I used my Claret glasses, which seemed to suit the wine very well. And the only time I use my *ballons* is when I have the luck to have a good bottle of Yquem, which does deserve the attention, for it can fill the entire space with its tremendous bouquet.

I have said nothing about smoking. All references to smoking in Holy Writ are unpleasant. [CAMPBELL: *This is one of the best and most appropriate* bons mots *in a book which teems with subtle wit, sly humour and provocative assertion, very Irish maybe, but widely entertaining and stimulating. Maurice's knowledge of wine in all its stages was, under the unusual circumstances, remarkable, and was picked up here, there and everywhere from books, letters, conversation and, of course, considerable personal acquaintance. I may have been critical of his judgments and conclusions, but that was only my taste and experience against his, and I may add that he was very obstinate once he had expressed an opinion. He was always rare good company, gay or serious as the spirit moved him, generous to a fault and prodigal of noble human sentiments which peep through his pages from first to last. His friends were the poorer when Maurice Healy untimely left them, but he will still live in the affectionate memory of all.*] Until recently I was in the habit of excusing my own abstention from tobacco by explaining that I should probably do my smoking in the next world. I have abandoned that jest since a friend replied to it the other day by saying: 'Ah! *Then* the fat *will* be in the fire!' But although I do not like the taste of tobacco, I do not object to its odour, in the proper place. When somebody smokes in a non-smoking carriage it is his rudeness and not his tobacco that offends me, and I like my own guests to smoke when they desire such a pleasure. I prefer that they should not do it in the dining-room itself, even after dinner; and I have a horror of the guest who produces a cigarette-case as soon as the joint has been removed. I think that motorists and smokers are apt to be selfish in not recognizing that their own enjoyment can occasionally interfere with the enjoyment of other people. In particular, I resent the smoker who holds his lighted cigarette under your nose, so that you get all the fumes, while he makes no use of it. And I cannot stand the person who leaves a lighted cigarette on the edge of a table or shelf or other piece of furniture, and then forgets it. Subject to these comments, I am not hostile to tobacco. But it should never be used with wine; and although Sherry can stand up to it, I am firmly of the opinion that you will enjoy your Sherry all the better if you refrain from smoking as you taste it.

So now I bid you farewell. May a day soon come when we can once more enjoy our food and drink with no sense of greediness and no regard to rations. May we in the meantime repeat: 'The Lord has given; the Lord has taken away; blessed be the name of the Lord.' And just as the patriarch was spared through all his sufferings to reach a new prosperity and happiness greater than that which he saw perish, so may we too, with hearts that are grateful for the favours of the past, look forward with courage, with confidence, but above all with patience to those coming days when once more we may stay ourselves with flagons, and rejoice in the good fellowship that only wine can bring to us.

APPENDIX

THE LIQUOR OF BRAY
THE CHÂTEAU MILLE-SECOUSSES
(*Reprinted by kind permission of the Editor of 'Wine and Food'*)

THE LIQUOR OF BRAY

(*Written for the Tenth Anniversary of the Opening of the Hind's Head*)

When George the Fifth was still our King
  A knowing friend of mine said:
'You cannot do a better thing
  Than join me at the Hind's Head.
Let Tio Pepe start the feast
  And Hine's Old Cognac crown it;
The food between will take at least
  A dozen flasks to drown it.
For this is truth, that I'll maintain
  Until my dying day, Sir:
That ne'er was wine, in any reign,
  As good as the Liquor of Bray, Sir.'

Some people have Teutonic tastes;
  It well may be that those'll
Prefer to line their portly waists
  With perfumed Hock or Mosel.
Let Rauenthaler crown their glass,
  Or, if they prefer the option,
They can from their drop of Sherry pass
  To Piesporter Goldtröpfchen.[1]
For this is truth, that I'll maintain
  Until my dying day, Sir:
That ne'er was wine, in any reign,
  As good as the Liquor of Bray, Sir.

[1] I know, I know! this is meant to be a humorous poem.

If Burgundy be your delight
   Explain to me what dastard
Contrived to sin against the light
   And call this Montrachet 'Bastard.'
Or if it be red wine we pour
   I set the choice between ye
Of noble Richebourg Nineteen-Four
   Or Nineteen-Seven Musigny.
For this is truth, that I'll maintain
   Until my dying day, Sir:
That ne'er was wine, in any reign,
   As good as the Liquor of Bray, Sir.

But if your meal be one of state,
   Or special jubilation,
With Krug '28 beside your plate
   You'll rise to the occasion.
If Mumm's the word, let it be heard,
   Lest thirst should make the bride sick;
On honeymoon you call the tune
   If you pay the Piper-Heidsieck.
For this is truth, that I'll maintain
   Until my dying day, Sir:
That ne'er was wine in any reign,
   As good as the Liquor of Bray, Sir.

Within so English a resort
   You'll find that the most sober
Will take his share of a bottle of Port
   Till April, since October.
And here be Ports to tempt from the sea
   The most persistent sailor.
The Cockburn and Croft will do for me,
   And you'll have the excellent Taylor.
For this is truth, that I'll maintain,
   Until my dying day, Sir:
That ne'er was wine, in any reign,
   As good as the Liquor of Bray, Sir.

But wise men know the glass to choose,
   And I my fealty owe, Sir,
(Though other wines I shan't refuse!)
   To the Red Wine of Bordeaux, Sir.

And so I deem it only meet
  To dedicate my poem
To the 1864 Lafite,
  Drunk from a Jeroboam.
For this is truth, that I'll maintain
  Until my dying day, Sir:
That ne'er was wine, in any reign,
  As good as the Liquor of Bray, Sir.

The souls of the just shall rise to Heaven;
  And, 'mid the angel chorus,
I hope that to us it may be given
  To find our good wine before us.
While, if we sometimes damned to Hell
  Our corked or doubtful bottles,
They *might* reach a few of us there as well;
  Right glad to wet our throttles!
So this is truth, that I'll maintain
  Until my dying day, Sir:
That ne'er was wine, in any reign,
  As good as the Liquor of Bray, Sir.

THE CHÂTEAU MILLE-SECOUSSES

(*To be sung to the air of 'The British Grenadiers'*)

Some talk of Cos d'Estournel
  And some of Desmirail
(Yet fill their glass diurnal
  With English dinner-ale).
But I'll commend another wine
  For banishing 'la frousse'
With the glou-glou-glou-glou-glou-glou-glou
  Of the Château Mille-Secousses.

There really is a château
  So named, of modest rank;
You reach, by local *bateau*,
  Gironde's less-favoured bank.
To drinking first and second growths
  You quickly call a truce
When you hear the glou-glou-glou-glou-glou
  Of the Château Mille-Secousses.

There came to Bourg one autumn
  A caterer of note;
He saw *barriques*, and bought 'em,
  And shipped 'em on his boat.
And all Great Western Dining-Cars
  Were furnished with the juice
And the glou-glou-glou-glou-glou-glou-glou
  Of the Château Mille-Secousses.

So here's to the Great Western,
  And also to its wines;
It may not be the best-run[1]
  Of all the British Lines.
But it cracked a joke with every flask
  It opened for our use
Of the wine of the Château Un-Deux-Trois—
  Jusqu'à-Cinquante-Mille-Secousses.

---

[1] See note on p. 247.

INDEX

C

G

H

T